## Ram Publications

*Complete VLF-TR Metal Detector Handbook (The)*
Thoroughly explains VLF/TR metal/mineral detectors and
HOW TO USE them. Compares VLF/TR's with all other types.

*Detector Owner's Field Manual*
Explains total capabilities and HOW TO USE procedures of all types of
metal detectors.

*Electronic Prospecting (Revised International Version)*
Learn how to find gold and silver veins, pockets, and nuggets using easy
electronic metal detector methods.

*Gold Panning Is Easy*
This excellent field guide shows you how to FIND and PAN gold as quickly
and easily as a professional.

*Modern Metal Detectors*
This advanced handbook for home, field, and classroom study gives the
expertise you need for success in any metal detecting situation, hobby or
professional, and increases your understanding of all fields of metal detector use.

*Professional Treasure Hunter*
Discover how to succeed with PROFESSIONAL METHODS, PERSISTENCE,
and HARD WORK.

*Robert Marx: Quest for Treasure*
The exciting, almost unbelievable true account of the discovery and salvage
of the Spanish treasure galleon, *Nuestra Señora de la Maravilla,* lost at sea,
January 1656.

*Successful Coin Hunting*
The world's most authoritative guide to FINDING VALUABLE COINS with
all types of metal detectors. The name speaks for itself!

*Treasure Hunter's Manual #6*
Quickly guides the inexperienced beginner through the mysteries of FULL
TIME TREASURE HUNTING.

*Treasure Hunter's Manual #7*
The classic! THE book on professional methods of RESEARCH, RECOVERY,
and DISPOSITION of treasures found.

*Treasure Hunting Pays Off!*
An excellent introduction to all facets of treasure hunting.

*Distributed By*
**Gem Guides Book Co.**
3677 San Gabriel Parkway
Pico Rivera, Calif. 90660

# *Electronic Prospecting*

### *Revised International Version*

by

## Charles Garrett

## Roy Lagal

Ram Publishing Company/Dallas, Texas

# 20th Century Gold Rush ...

A RAM GUIDEBOOK

ISBN 0-915920-38-7
*Library of Congress Catalog Card No. 76-11380*
*Electronic Prospecting*
© Copyright 1979. © Copyright 1980. © 1985 Charles Garrett, and Roy Lagal
First printing February 1979. Seventh printing June 1986.

ON THE FRONT COVER: Roy Lagal, the father of today's new electronic prospecting methods, holds a double handful of large gold nuggets worth many thousands of dollars. Patience and the new VLF (Very Low Frequency) electronic detection system now make it possible even for beginners to find nuggets as small as one pennyweight, with ease.

# CONTENTS

The authors wish to thank Bob Grant for his assistance in editing and compiling our notes and operational methods and techniques into organized book form.

# MEET THE AUTHORS

CHARLES GARRETT, President of Garrett Electronics, Inc., has been treasure hunting and prospecting since his teenage years. He has traveled in Australia, Canada, Europe, Mexico, and in the United States in his search for treasure and precise personal knowledge of detector applications in order to perfect not only his treasure hunting techniques, but the design of the metal detectors that he manufacturers. Charles has been writing for the treasure hunting and prospecting field since the late 1960s. In addition to numerous articles, he has written several books, including the popular SUCCESSFUL COIN HUNTING, MODERN METAL DETECTORS, and co-authored, with Roy Lagal, several others. His field tested and proved metal detectors have greatly increased the capabilities and success of prospectors the world over and have caused, in great measure, this centuries electronic prospecting gold rush.

ROY LAGAL knows more about prospecting with the aid of electronic metal/mineral detectors than any other man. His knowledge and expertise are based upon a lifetime of treasure hunting and prospecting with metal detectors. Roy and his wife have traveled for more than forty years in the United States, Mexico, and Canada prospecting for gold. For nearly twenty years he has worked extensively with Charles Garrett and others to develop the field of electronic prospecting to a fine art. Roy, designer of the "Gravity Trap" gold pan, author of five books and co-author of two others, gives unselfishly of his time and knowledge about prospecting and treasure to all who ask. His COMPLETE VLF-TR METAL DETECTOR HANDBOOK, co-authored with Charles Garrett, along with ELECTRONIC PROSPECTING, and his latest book, WEEKEND PROSPECTING elevated the prospecting hobby to a professional level that anyone can quickly master.

The authors have received credits and acclaim from throughout the world for their books and articles written on electronic prospecting methods. As early as the late 1960s, Charles and Roy began dispelling the advertising untruths and myths that surrounded the use of metal detectors when employed in locating ore bodies and small nuggets of gold/silver. Combining advanced detection systems with countless hours of research, produced success. Charles Garrett traveled more in the United States and several foreign countries prospecting with electronic detection systems than all other manufacturers combined. Roy Lagal attempted to adapt electronic methods to prospecting long before the system itself could be regarded as successful. Their combined efforts led to the first authoritative books published on such methods and produced the new electronic gold rush of the 1970s. Literally tens of millions of dollars have been found by rank amateurs as a result of their published works. The successful searcher who has already found gold can attest to this.

# FOREWORD

ELECTRONIC PROSPECTING is *the* guidebook for the in-the-field prospector. There is no other book written exclusively for the prospector who wants to learn successful field searching techniques using the VLF/TR ground canceling metal detector. With the introduction of the VLF/TR metal detector, a whole new world has opened up. For instance, in addition to the detector's capability of detecting all conductive metals (gold, silver, copper, led, cinnabar, etc.) to super depths while canceling the effects of earth's minerals (ferrous iron), almost any irregularity in the earth's formation (iron dike, magnetic vein, mineral change, etc.) can be detected providing the operator understands how to adjust and operate the VLF/TR detector correctly.

The authors have instructed the reader in the basic techniques needed to be successful regardless of his or her brand of VLF/TR detector. The instructions and techniques presented herein are applicable in any given locale, be it Africa, Australia, Canada, Central or South America, Europe, the United States, or any other country. The VLF/TR metal detectors and their fantastic, unparalleled capabiities should be learned by all persons who search for the earth's most valued treasures.

Editor
*Western and Eastern Treasurers* Magazine Ray Kruja

# CHAPTER I

# The Lure of Gold

Since the time the first shining lump was pounded into an ornamental object, gold has been one of man's most valued, sought-after treasures. Although gold was too scarce in very early times to be used in daily transactions, it was used as a store of value by the ancient civilizations — in China from about 1200 B.C.; in Egypt from about 1000 B.C.; and in Babylon and Minoa from the third century B.C. Nearly two thousand years later the Spanish followed the lure of gold into the New World, plundering the civilizations of the Inca and the Aztec. More than two hundred years later the California gold rush epitomized the fervor of the modern search for the precious yellow metal that continues to this very day.

Even though countless tons of gold have been mined, the following brief study will show that most of the earth's gold is still waiting to be recovered. In 1849, the year of the big California gold rush, gold production was 11,866 ounces. Production climbed steadily, reaching a peak of 2,782,018 ounces in 1856. Nine years later production had declined to 867,405 ounces. The great California gold rush was over; the miners had moved on to greener pastures. However, even with the surface gold removed in the Mother Lode country of California, numerous geological surveys and studies have suggested that only some 15% or 20% of the gold in California has actually been recovered. Based on this data, it seems obvious that a vast amount of gold still remains to be discovered, not only in California but in all parts of the world.

## THE 20TH-CENTURY GOLD RUSH

On December 31, 1974, the United States government ended the ban on the private ownership of gold which had been inaugurated 41 years previously by President Franklin Roosevelt as part of his program for shoring up the nation's failing economy. The price of gold began to climb immediately and, even though there have been peaks and valleys, the market price of gold has steadily increased. This increase in value, coupled with man's pioneering spirit, has caused a small scale rush back to the 19th-century gold camps.

The rugged '49'ers who began their trek to the gold fields of the American West in 1849, later moving on to

1

The large 20-ounce gold nugget shown in these three photographs was found by Australian electronic prospector, Tom Murray of Laverton. Peter Bridge, Hesperian Detectors, P. O. Box 317, Victoria Park 6100, Western Australia, reports that Tom found the nugget in the gold fields north of Perth. The lower photograph on the opposite page depicts the nugget as it was being recovered. With the aid of ground canceling detectors thousands of ounces of gold nuggets have been found recently in the Australian gold fields. Electronic prospecting with these new type detectors is certain to open other (dry) areas that cannot be investigated by normal prospecting methods due to lack of water. In addition to finding the 20-ounce nugget, Tom found a lump of quartz containing 50 ounces of gold.

2

Australia and other countries, exemplified the true mood of modern man. The '49'ers wanted to make their own fortunes. That same spirit still prevails in the gold producing areas of the world today, where both professional and recreational prospectors have discovered the pleasures and profits of searching for gold.

## WHO IS PROSPECTING?

Doctors, attorneys, businessmen, students, retirees, entire families from all walks of life have found that the healthful, relaxed outdoor life of weekend or vacation prospecting can yield big dividends. A single ounce of gold is now, in many cases, worth several days' pay. One fair-sized gold nugget can easily be the equivalent of a month's salary. This is obviously part of the big attraction of the business-hobby called "recreational mining." The people searching for their own gold are people who are working at regular jobs, but, on weekends and during vacation time, they load the family into the car and head for one of the many thousands of gold areas open to the public. They may camp by a stream and spend their time panning or dredging for gold or silver, or they may travel to an old ghost town

This is a seven-pound slab of pure native silver and gold that was taken from a mine. Conductive veins of this nature, and even non-conductive, magnetic (highly ferrous) veins, can be easily detected several feet into mine walls and floors. Ground canceling detectors are not affected by uniform ground mineralization and can easily locate conductive metals and out-of-place magnetic ore deposits. Oftentimes, highly magnetic ore deposits will contain rich quantities of gold, silver, and other sought after metals.

4

or deserted mining camp and search for ore veins or valuable mineral/ore specimens overlooked by early day prospectors.

## MODERN EQUIPMENT

Three developments have greatly increased the ability of the recreational miner to hit paydirt: the availability of an easy-to-use, highly efficient gold pan; the production of extremely portable, lightweight, and efficient gold dredges; and, the development of the one tool that makes locating precious metals simpler, either on land or in the water — the rugged, highly sensitive ground canceling metal detector capable of operation and detection in even the most difficult mineralized soil or rock.

# CHAPTER II

# Where You Can Find Gold

While there is some gold in nearly every state of the Union, the precious yellow metal is present in sufficient quantities to make prospecting profitable in only about half of the states. In the other half, the gold is so fine and in such minute quantities as to make recovery impractical.

Although it is not judicious to prospect for gold in *every* state, the major gold producing states are *not* concentrated in the West, as may be thought at first, but they are, instead, liberally sprinkled around the entire country. The major gold producing states, roughly in order of their production, are listed below.

| | | |
|---|---|---|
| California | Arizona | South Carolina |
| Colorado | Oregon | Tennessee |
| South Dakota | Idaho | Virginia |
| Alaska | Washington | Alabama |
| Nevada | New Mexico | Texas |
| Utah | North Carolina | Michigan |
| Montana | Wyoming | Wisconsin |
| | Georgia | |

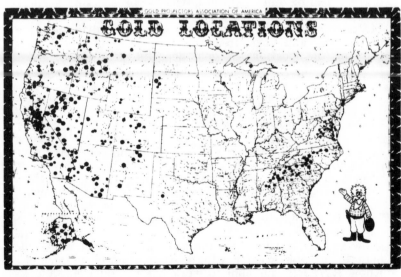

This map is included only as a guide to U.S. gold producing areas. See the appendix, dealer listing, and recommended books for sources of information (state bureaus of mines and geology, etc.) about gold locations in the U.S., Canada, Australia, and other countries.

The combined production of the gold producing states is more than 300,000,000 ounces. At today's gold prices this production is worth more than sixty billion dollars! As is evident, gold production in the United States has been no small matter, and in many states production continues today.

Take a close look at the map to locate the high production area nearest to you and do a little research. California earned the number one place, as shown on the above list, with a total production since 1799 of more than 100,000,000 ounces. South Dakota earned third spot with a production of 31,000,000 ounces; Arizona, eighth with 13,000,000 ounces; and North Carolina, thirteenth (not all that bad!), with a total production of 1,000,000 ounces.

## DO SOME INVESTIGATING

Once you have decided upon the area(s) in which you wish to prospect, check with the governmental agencies and request literature that would be of assistance to you. All of the states with "public lands" are under Bureau of Land Management (BLM) regional offices (this includes most of the western states), and the BLM has a number of helpful pamphlets on staking claims and other matters which affect the prospector. Many states have a Bureau of Mines, and most of those offices have literature showing gold producing areas. The Bureaus of Mines can also offer a great deal of other helpful advice on prospecting geology and gold recovery. A third place to write is the appropriate State Department of Tourism. Many states realize the value of tourists, and have prepared maps and other materials showing the locations of gold deposits, gem fields, ghost towns, mining districts, and other points of interest to visitors.

Go to the public libraries. Check out as many books as you can on the gold districts you are planning to work. Find out what type of prospecting there is to do. Are there dredge tailing piles to work? Are there good placer areas (see Chapter XI) where you can find nuggets? Are you going to search for rich gold ore veins? Decide *precisely* what type of prospecting you are going to do, and then equip yourself accordingly. If possible, talk to other prospectors who have worked the areas you intend to enter. There is absolutely no question about the fact that the more familiar you are with the area and the more you know and are equipped for the particular type of work you plan to do, the greater will be your chances of success. Know *WHERE* you plan to work before you ever take to the field, and there is little doubt that you will come back with some good specimens of gold!

Chuck Blem yells with glee as George Massie (right) and a Gold Prospectors Association of America (GPAA) club member inspect gold they panned from a stream in Northern California. The lower photograph shows the results of a few days' work! Note the large nuggets! George Massie is founder and president of the GPAA. The organization conducts continuous instruction seminars throughout the United States to teach modern-day panning methods and demonstrate the latest equipment for the recreational miner, electronic prospector, and treasure hunter. One piece of prospecting equipment demonstrated, used by, and recommended by GPAA is the "Gravity Trap" gold pan manufactured by Garrett Electronics. The GPAA sponsors numerous state and national contests. Contact George Massie for further information.

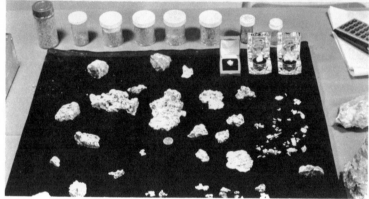

Two of the authors, Charles Garrett and Roy Lagal, operate their treasure booth at one of the numerous western U. S. gold shows. The large gold nuggets and other valuable specimens in the collection were found in the United States, Mexico, Australia, Canada, and Alaska by several different operators using VLF/TR electronic prospecting methods. Charles and Roy have hunted gold in several of the western and northwestern states where they have found gold nuggets ranging in size up to two-and-one-half-pounds. Many of the silver specimens were found by Charles in Mexico. Extremely valuable treasure collections such as this one are being amassed much more quickly because, over the years, these two men had faith in electronic prospecting and continued their efforts to perfect metal detector design and usage, thus giving treasure hunters the long awaited equipment. In fact, today's electronic gold rush would never have been possible without the development of correctly designed and calibrated VLF/TR ground canceling metal detectors.

## YOU CAN GO TO SCHOOL

George Massie, President of Gold Prospectors Association of America (GPAA) schedules approximately fifty prospecting sessions each year throughout the United States. He travels to the major cities, presenting the latest gold recovery techniques to those persons interested in this fascinating field. The GPAA publishes a periodical that is loaded with facts and figures. Those who are interested in becoming a member of GPAA may write to George Massie, GPAA, P. O. Box 507, Bonsall, California 92003.

Those of you who are unable to attend such "hands-on" events should investigate the "how-to"books published by Ram Publishing Company, Dallas, Texas. Such books are easily read at leisure and often result in some amazing discoveries. Probably ninety percent of today's "weekend prospectors" began with these instructive guides and, as a result, most are highly successful.

# CHAPTER III

# Various Recovery Methods

When searching for gold deposits, you will find some areas are naturally going to produce more color than others. Remember that gold is a heavy metal and will tend to follow the shortest path as it moves along a stream or river. As you look down a stream having a long stretch of fast moving water, keep alert for cracks, crevices, or other anomalies on the bottom which will stop the movement of the gold and trap it. At the end of long stretches, search for gold deposits at the *inside* of bends in the river.

One thing that must always be remembered about searching for gold is: gold is where you find it! Although many obvious locations have produced great amounts of gold in the past and may still continue to do so today, conditions over hundreds and thousands of years may have left large deposits of gold in places that today look highly unlikely.

## THE GOLD PAN AND YOU

After you have done your research and located a good gold area with the aid of your metal detector or when you have turned up some "color" in a pan full of river gravel, you must then separate the gold from the other materials. Panning is the most popular method of retrieving the gold for a number of reasons: pans are inexpensive, easy to carry, and, with a little knowledge and practice on your part, easy to use. If you are near a stream or other water source, or if you can rig up a wash tub with sand, gravel, and water, learning to pan is simple. Dry panning, though a little harder and a bit more dusty, is still relatively easy to learn.

Regardless of whether you intend to recover gold by searching for it with a metal detector, by dredging, or by sluicing, the gold pan will remain one of your primary tools. Taking a little time at the beginning to familiarize yourself with the most efficient techniques of panning will pay big dividends in the field. In addition to the normal uses of the gold pan, the pan may also be used for easy recovery of metallic targets signaled by your detector, which objects would otherwise be difficult to locate. More on that later. . . .

During the gold rush days of the late 1800's and the early 1900's, gold panning was back-breaking labor and, unless the early day panners hit a lucky spot, it was usually

Plebe Ciro of Liverno, Italy uses a Garrett gold pan to investigate a response found with his VLF metal detector. Black sand deposits may be located in the same nugget searching procedure, if desired. Simply increase the manual ground canceling until the ground responds as metallic. Generally one or more full turns into the positive zone on the detector easily accomplishes this. A constant operating height should be maintained and all indications investigated. Nuggets, magnetic black sand concentrations, and "hot rocks" will all respond the same... metallic. Bandits often frequent this mountain area and Plebe, along with the photographer did not remain too long. Remember, this dual method of detection is one of the best but, the least known or used. Try it.

not very profitable. Today, however, gold panning is a lot easier, and a great deal more material of value is being recovered for the amount of work performed. This is due not only to the greatly increased price of gold on the open market but also to the new, efficient, green "Gravity Trap" gold pan.

The "Gravity Trap" gold pan was designed by Roy Lagal of Lewiston, Idaho, and is manufactured and distributed by Garrett Electronics, Garland, Texas. Made of light but durable ABS plastic, the pan weighs much less than the old style metal pans. More importantly, the "Gravity Trap" gold pan has built-in gold traps in the form of sharp 90-degree riffles. These riffles trap the heavier gold and allow fast panning off of the unwanted rocks and gravels. The pan is a forest green color which has been proved in extensive laboratory and field tests to show gold, garnets, precious gems, and black sand better than other colors, including black. With a little practice, a weekend or recreational placer miner using this new pan can work with equal or greater efficiency than the most proficient professional using the old style metal pans or those of plain black plastic.

Today's recreational miner can achieve excellent results by using a good ground canceling type metal detector to locate gold deposits in conjunction with panning the promising locations with a "Gravity Trap" gold pan. Not everyone who hunts gold is going to get rich. It is safe to say, however, that weekenders and vacationers, if they pursue placer (pronounced "plaster" without the "t") mining properly, can offset a goodly portion of their expenses by panning gold. Equally satisfying is the pleasure of sitting by a clear mountain stream or a long-forgotten dry gulch, producing income with your own two hands, knowing full well in the back of your mind that there is always the chance of hitting that "big one"!

With the "Gravity Trap" gold pan, even dry stream beds and sand washes can be made to produce gold. This pan allows both wet and dry panning since the built-in riffle design can be depended upon to trap the gold. In fact, dry streams which have not seen water for many years can be very productive because they probably were passed by during the busier gold rush days. The old-timers, with their less efficient metal pans, preferred to work with running water because it made panning a lot easier. However, with today's improved pans and electronic detection equipment, new gold producing areas are being discovered daily, and the known producing areas of the past are giving up the gold deposits the old-timers missed. The fun, excitement,

Virgil Hutton, Tommie T. Long, and co-author Charles Garrett work a remote area on the famous Salmon River in Idaho. Three or four members of the crew makes it possible to pack complete small dredging equipment and other prospecting tools into such remote areas. Single prospectors should use the back-pack method and carry only the most productive equipment, a lightweight plastic gold pan, and a good VLF detector with a manual ground control. Remember, not ALL detectors are suitable for prospecting.

and profit of recreational mining are out there waiting in the beautiful gold country. The treasure hunters of today are limited only by their time and desire.

## LIGHTWEIGHT GOLD DREDGES FOR THE WEEKEND PROSPECTOR

Gold dredges, crude and often as big as a full-sized house, have been used for gold recovery since before the turn of the century. Dredging is by far the easiest way to retrieve gold from river and stream gravels. The modern, lightweight suction dredges of today are capable of working bedrock cracks and crevices like an underwater vacuum cleaner.

Powered by a small gasoline engine which drives a centrifugal pump used to create the vacuum needed to suck up sand, rock, gravel, and other gold-bearing materials from the stream bed, modern dredges employed by the recreational miner are extremely efficient and light enough to be carried easily by one man. The sluice box of the dredge is designed to retain even the fine gold, yet spill large rocks and other, lighter-weight materials into the tailing pile without clogging the riffles. The relative ease of cleanup depends upon the size and type of dredge used.

A. M. Van Fossen, Charles Garrett, and L. L. "Abe" Lincoln fire up a lightweight gold dredge in the Sierra Madre Mountains in Mexico. The small type dredge was selected because the men had to hike eighteen miles down river to this isolated but rich ore-bearing location. The VLF/TR metal detector proved reasonably successful in locating black sand deposits. These concentrations generally contained the most gold, as the magnetic sand is heavier than the beach sand and tends to concentrate much the same as the heavier gold. Any type VLF/TR detector may be used for this application. Simply adjust the ground canceling control in VLF mode until the soil, gravel, and rock responds as positive (metal); maintain a uniform operating height; and scan the area, either underwater or above, depending on depth. The heavier concentration of mineral will respond louder. So will a large gold nugget, simple but effective.

Dredge operation is simple. The suction hose (which ranges from 1½ inches all the way up to about 6 inches in diameter on the larger recreational dredges) is worked slowly along, chewing away the gravel bank, reaching ever deeper for bedrock. Color in the auriferous (gold bearing) gravel is recovered, but the richest deposits are usually trapped in crevices in the bedrock itself. The suction is sufficient, along with the use of a "crevicing tool," to break up the jams in the cracks and to clean the rich deposits of fine gold and nuggets out of the crevices. The material is pulled up through the hose and dumped onto the riffle box. The flow of water keeps the sand, rocks, and lighter materials moving along the riffles until they drop out of the opposite end of the box into the water. The heavier gold flakes and nuggets are trapped behind the riffles where they remain until the concentrates are panned down, usually at the end of the day.

Before you actually use a suction dredge in a river or stream, be sure to check with local authorities. Some states, like California, require a dredging permit obtainable from the local office of the Department of Fish and Game for a fee of $5. Other states actually prohibit dredge operation in certain locations. Some of the laws came about in the distant past because a few irresponsible, large-scale dredge operators left ugly scars on the countryside. An unfortunate situation is sometimes created because the small dredges in use by the recreational miner today do not damage the ecology since all material is returned to the streambed. In fact, the turn-over of the stream bottom actually provides new supplies of food for the fish, thereby helping the ecology. In the older dredging operations, the large equipment piled huge stacks of rocks alongside the stream, producing not only an eyesore but also, in some cases, actually changing the water course.

Keep in mind, however, if you encounter one of these old dredge piles, that there is probably some good gold to be found with your metal detector. Oftentimes solid gold nuggets and gold enclosed inside large, everyday-appearing rocks passed right through the dredges and were dumped out with the tailings. Your ground canceling metal detector can easily locate this gold missed by the early day prospectors.

CHAPTER IV

# Can Gold Be Found with a Metal Detector?

"Can I find gold with a metal detector?" This is the question most often asked by gold seekers who are interested in improving their chances of success but who have doubts about their ability to find gold electronically. Many persons have been misled into thinking they can rush out into gold producing areas with *any* type detector and find gold. (This impression may have been created by several metal detector manufacturers who have been overly enthusiastic in claims for the capabilities of their instruments.) Until recently, the Beat Frequency Oscillator (BFO) detector was the only type that could be depended upon to produce true, accurate readings. The BFO, however, is limited because of its marginal depth penetration in mineralized ground.

Today's modern, ground canceling detectors, of which the VLF/TR is a type, offer greatly expanded capabilities: ground mineral problems have been mostly overcome; detection depth has been tremendously increased; and, as a result of extensive field and laboratory tests and careful electronic design and manufacturing techniques, detectors capable of very exacting metal/mineral locating and identifying characteristics are now being built.

Using VLF/TR discriminating detectors, both professional and weekend prospectors are making rich strikes over previously worked areas, unearthing nuggets similar to those found at the turn of the century. Among the gold and silver strikes made with detectors are the following.

- In central Washington Roy Lagal found a 2½-pound gold nugget.
- A West Coast metal detector manufacturer reports that one of his customers found a beautiful, 36-ounce gold nugget near the central Cascades in Washington.
- Charles Garrett spent several weeks prospecting in Mexico and in one mine alone found two silver pockets and one silver vein which was quite large.
- A 47-ounce nugget found at Mt. Magnet, Western Australia, was valued at $7,000 for its gold content but at $25,000 as a specimen.
- A 25-ounce nugget and a 36-ounce nugget were found with VLF/TR detectors in the northeastern Australian gold fields.

19

- Peter Bridge of Victoria Park, Western Australia, reports that more than 100 pounds of gold nuggets have recently been found by his customers using the new VLF/TR metal detectors.
- Two-hundred ounces of small nuggets were recovered by two prospectors in Western Australia. One detector operator, working alone, has recently found more than 500 ounces of large gold nuggets.
- In his store in Houston, Texas, A. M. VanFossen proudly displays many exquisite native silver specimens which he recovered in one of the seven Chihuahua, Mexico, mining districts.
- After a gold panning weekend in Georgia, one lady treasure hunter returned home with a vial of beautiful, tiny gold nuggets.

## AUSTRALIAN GOLD

When one man, equipped with a VLF/TR ground canceling metal detector, can find more than five-hundred ounces of gold in a few short months, it is time to sit up and take notice. A small scale recreational mining boom has suddenly taken place in Australia, and it behooves anyone interested to check into the situation further. Peter Bridge, one man who owns and operates a prospectors' supply house in Western Australia, has been largely responsible for introducing the VLF/TR metal detector in Australia. Those who wish to learn more about this fascinating gold situation should contact him at Hesperian Detectors, Box 317, Victoria Park 6100, Western Australia, (09) 38-66105.

Although no official figures are available, the SUNDAY TIMES of Perth, Australia, reported, "It is estimated that more than 2,000 ounces of alluvial gold has been picked up by Western Australian prospectors in the past two years. The value of these finds exceeds one-half-million dollars!"

This new Australian gold rush is attributed to two major factors: (1) banking (gold) regulations were recently repealed and gold can now be sold on the open market in Australia; and (2) sophisticated VLF/TR discriminating metal detectors which give prospectors a highly efficient gold-finding tool have been introduced!

## IN LIBERTY

A little closer to home, Charles Garrett and Roy Lagal joined Frank Duval at Liberty, Washington, to search the old dredge tailings, conveyor piles, and mine pits for nuggets and rich ore deposits. That area of the country is very highly mineralized, but the results produced by the new VLF/TR type detectors are truly amazing.

"Mining in this district has been described as different from any other mining district in the world," Roy Mayo writes in his book, GIVE ME LIBERTY. "Placer gold has been found as water worn, rounded nuggets, indicating travel over a long distance. Nuggets have also been found having sharp edges with pieces of quartz still attached, indicating nearness to their source. Flakes of gold and delicate wire gold are also found. This is one of the few places in the world where gold may be found in its crystalline form."

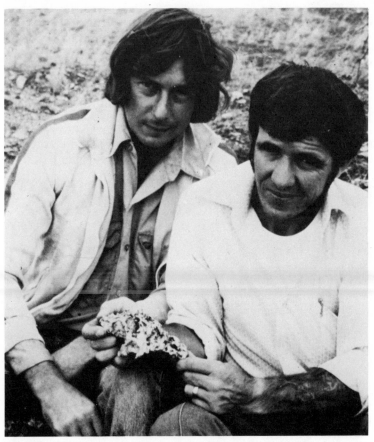

Stewart Hooper and Noel Scattini proudly display this beautiful, unique gold nugget which they found during their electronic search for gold in Western Australia's Outback region. This 56-ounce nugget, valued at more than $75,000, proves that it is possible for today's outdoor adventurer to strike it rich quickly. This nugget is representative of thousands of gold nuggets which have been and are being found in Australia and other countries of the world. The soil in this particular area contains extremely heavy concentrations of iron minerals. This find, and others like it, would be impossible without today's VLF/TR ground canceling metal detectors.

Lode gold is also found around Liberty. Shale and clay pockets contain delicate wire gold, and gold can also be found in stringers in the sandstone and in a quartz and shale mixture called bird's-eye quartz. The major discoveries of wire gold have been on Flag Mountain and Snowshoe Ridge, according to Mayo. Many beautiful specimens of wire gold weighing several pounds have been found there. Anyone wishing to learn more about Liberty may order Mayo's book by writing to Mayo Maps, P. O. Box 184, Enumclaw, Washington 98022.

The purpose of our trip to Liberty was to do some electronic prospecting. That area was chosen because of the wide variety it offers both of types of gold deposits and types of hunting. We had the opportunity to work the old dredge tailings along Swauk Creek where, over the years, picknickers have found many large nuggets. We also had the opportunity to check for pockets of wire gold on Mrs. Bertha Benson's claim on Flag Mountain and to pinpoint both non-magnetic conductive ore veins and magnetic non-conductive ore veins. All these activities were coordinated by Jacob Kirsch who has been mining his Liberty claims for many, many years.

This arrowhead-shaped gold nugget was found by Roy Lagal in old dredge tailings near Liberty, Washington. Careful searching with the newer VLF/TR type detectors in areas where older style dredging equipment was once used can sometimes be very profitable. Early-day dredge operators were unaware of the presence of large nuggets and they were passed into the tailings pile. Today, they miss very little, if any gold. Practically all large scale operators inspect the bare bedrock after removing the gravel by machinery, and before restoring or replacing the gravel to its original site. It's a proven fact that this inspection often discovers more gold nuggets embedded in the undisturbed cracks of bedrock than was recovered from the processed gravel. Those who do not, are leaving the gold that could be recovered at the least expense.

Working the dredge piles along Swauk Creek was a real eye-opener. Setting the ground canceling control to cancel out the normal ground mineralization, we began sweeping our detectors slowly over the large boulders, searching for elusive gold nuggets that might have rolled off the trommel many years ago while the dredge was working. Apparently the dredge used a trommel (a sieve for cleaning or sizing of ore, coal, *etc.*) with small holes, and many nuggets larger than the openings in the screen were reclaimed by the creek as they passed over the screen and were diverted into the tailings piles.

As we worked our way along, we really discovered the true meaning of "hot rocks," those out-of-place, highly mineralized rocks that can easily fool an electronic prospector if he is not aware of what they are all about. Charles Garrett and Roy Lagal showed how easy it is to identify hot rocks quickly using the ground canceling detector's TR discriminate mode. (The subject of hot rocks and how to identify them with the discriminator circuit of your detector will be covered in more detail later.)

While it is often necessary to cover a good bit of ground and move a considerable amount of gravel, the end results of prospecting can be well worth the trouble. A couple of hours' work with detectors and spades produced a beautiful arrowhead-shaped gold nugget for Roy Lagal!

On the hill above Liberty, Charles Garrett and Roy Lagal carefully searched for pockets of gold and highly mineralized veins on a rich hardrock claim that has produced many thousands of dollars in gold. A number of veins were discovered and mapped out to be excavated and examined for precious metals at a later date. These findings proved beyond any doubt that the new type ground canceling metal detector is an invaluable aid to the prospector, the most valuable tool a weekend miner or professional prospector can own. Properly used, the detector can find conductive (non-ferrous) and non-conductive (predominantly magnetic) deposits, as well as nuggets, pockets, and veins of almost any nature, providing sufficient quantities of magnetic iron are present.

Charles Garrett and Roy Lagal (facing camera) are supervising an exploratory expedition into one of the rich silver districts of Old Mexico. Located in the Canyon de Cobre region of Mexico, the area contained little gold values but was mined for silver by the Spanish and later by more modern methods. Little of value was recovered by the small dredge due to the fact that the fine silver was closely related in weight to the magnetic black sand. However, huge silver speci- mens weighing several hundred pounds were discovered using the VLF/TR detectors along and near the river's channel. Also in the background can be seen portions of the ancient smelter. Worked by slave labor and inefficient methods, they recovered several "spills" of pure silver lost during smelting operations. Here, a beginner using a VLF/TR detector could recover his weight in silver easily, but bringing it out of Mexico would be quite difficult. The natives quickly adapted the Garrett "Gravity Trap" gold pan to their own methods of recover- ing the fine silver found in the rivers.

24

CHAPTER V

# The Ground Canceling "Seeing Eye" Metal Detector as a Tool

Only within the last decade or so has the electronic metal detector come into its own as a prospecting tool. Sensitive, lightweight, stable, and reliable, today's metal detector serves well as an extra pair of "seeing eyes" for the modern prospector. The prospectors and miners who took part in the gold rushes of the 1800's had only one tool with which to detect gold — their eyesight. If they didn't see the gold itself or recognize gold-bearing rocks, they could not come out with any color or nuggets.

Gold can be found in a number of forms, among which are: lode or hard rock deposits found in a vein and often mixed with other materials; placer deposits, either in a stream or in dry sands and gravel; and nuggets of the pure metal. (Nuggets can also be found in vein material or as part of a placer deposit.) Placer gold is the type most often sought after by the weekend miner, but more and more treasure hunters are taking their detectors into the mountains in search of rich lode deposits.

Placer deposits are generally formed by weather erosion of an outcropping of lode gold. As the outcropping is decomposed, water carries the material down into streams where the large chunks are ground into sand and gravel, thus releasing the gold from the lode. Due to its extreme weight, the gold works its way down to the stream bedrock where it is trapped in cracks or crevices while the lighter sand and gravel are swept on by the rushing waters.

While most placer gold was originally deposited by water, many good placer deposits today are high and dry. Sand washes in gold producing country will many times contain a high concentration of fine or so-called flour gold that can be recovered easily either by dry panning or by use of a simply constructed dry washer. Many times, mixed in with the finer gold will be small nuggets that can be detected easily with a good metal detector.

Working vein or lode gold may require a bit more experience and research if the weekend miner is to be successful. While a smattering of knowledge about geology can be most helpful, the treasure hunter can do well simply by working around abandoned mines and mine dumps. Many times, a miner would miss a rich ore vein by mere inches and give up, thinking the gold had played out. Sometimes, with only their eyesight to guide them, the

25

miners of the 1800's carefully followed a vein of gold deep into a mountainside, yet they could have been only inches away from a vein incredibly richer than the one they were following. Your metal detector, properly worked along the walls of mine tunnels, may just show you the precise location of that incredibly rich vein.

While prospecting with his metal detector in Chihuahua, Mexico, Charles Garrett located three silver veins that had been missed by the early day silver miners. In scanning the walls of the mine owned by Javier Castellanos, Charles received three readings that necessitated further investigation. Two of the readings proved to be silver pockets, whereas the third proved to be a vein of native silver. The vein was detected in the area where it was one-half-inch wide. As Javier tooled his way into the mine wall, the vein began to enlarge as it continued down past floor level where it became several inches wide. The pockets and the vein perhaps would have been lost forever had not the electronic metal detector discovered their presence.

Don't overlook the goodly supply of ore that is always lying around on the tunnel floor or beside the rails on the way to the mill. By using a detector on these ore samples, you can sometimes uncover many good specimens which previous miners missed. When a sample containing no visible gold would fall off the ore cart, it often would be ignored.

Dumps and tailing piles of old mines are good locations to work. They are certainly the easiest and quickest sources of gold if the prospector knows how to use a metal detector. Many times large gold or silver nuggets are concealed inside a chunk of rock and are unknowingly discarded on the mine dump. While the old-time prospector couldn't see the valuables inside the rock, your metal detector certainly can! So don't overlook working the discards in the tailing pile around abandoned mines. Many times more gold is still in the ground than was recovered in all the California, Colorado, and Alaskan gold rushes combined. It is yours for the taking!

# CHAPTER VI

# The Plastic Gold Pan: Companion to the Metal Detector

The metal detector and a good plastic gold pan are the dynamic duo of electronic prospecting. The metal detector can be of great aid in locating nugget deposits and good panning locations. If your search is for large nuggets and the location is a highly mineralized creek bed that abounds with rocks of high mineral content, the VLF/TR discriminator detector will be of outstanding assistance. By adjusting the ground canceling control of the detector to eliminate the highly mineralized sands, there will be super-sensitivity to most small conductive gold nuggets and definitely to the larger ones.

When you are conducting an electronic search, a plastic gold pan can make target recovery much easier. Whether the target turns out to be only a spent bullet or other metallic object in the stream, or whether it turns out to be a large gold nugget, proper use of the plastic gold pan will save a lot of recovery time.

Roy Lagal's detector has just sounded off with a metallic response. That the target was metallic was determined by switching the ground canceling detector into the true TR discrimination mode. Roy is shoveling the rocks and pebbles into the plastic gold pan. He will then scan the pan with his detector. This is often the most practical way to locate detected objects because the eye cannot see conductive gold inside a rock.

The technique is simple. First, be certain to pay close attention to faint detector signals as they may indicate the nuggets are rather small or that they are deeply buried. When you receive a target response with the metal detector, pinpoint it as closely as you can. Then slip a shovel under the spot from which the signal came. Be extremely careful as small, heavy metallic objects always tend to sink into the gravel and thus may be lost. Place the small amount of gravel or sand you scoop up into the plastic gold pan and check the material with your detector. If the target is in the pan, the detector will respond. If there is no response, dump the pan and try to get under the target on the next attempt. With a bit of practice you will become quite proficient in this technique (even though you may be working in water three feet deep!) and you will soon discover that it will save a lot of time.

Needless to say, the *plastic* pan is a necessity for this procedure. Obviously, you cannot use a *metal* pan for checking material with your detector! You will quickly find that

Charles Garrett and Roy Lagal check the silver (about 5 pennyweight) that the Mexican prospector panned from one shovel full of sand and gravel taken from bedrock. Mexican prospectors instantly recognized the value of the patented 90-degree riffle "Gravity Trap" gold pan and the VLF/TR metal detectors when the authors demonstrated the capabilities of the equipment. This sharing of knowledge and expertise is a major reason why treasure hunting and electronic prospecting have rapidly come to be a world-wide activity which is resulting in the recovery of millions of dollars worth of precious metal that might otherwise remain hidden forever.

use of the plastic pan with the built-in "Gravity Trap" principle provides not only a simple, efficient method of sorting through the sand and gravel to locate the metallic object that caused your detector to respond but also a quick way to settle *ALL* the heavier concentrates and separate the lighter material.

The recovery technique is the same when searching old dry wash or placer diggin's, except the object will be easier to locate than it will be in a stream. Old dredge tailings can be a bit more difficult to work, and you will probably lose many metallic objects on your first tries. The material is loose and the heavier metallic objects easily work their way down through the tailings. Once lost, they are very difficult to recover as they work their way deeper with each attempt to find them and become nearly impossible to relocate. On the other hand, you'll be surprised at how quickly you become proficient at this technique once you realize that those metal objects you are losing just might be gold nuggets.

# CHAPTER VII

# Low Frequency VLF/TR
# Identification of Metal vs. Mineral

Provided they are in a conductive form and in sufficient quantity to disturb the electromagnetic field of the search-coil, gold, silver, copper, and other valued non-ferrous metals will produce a *metallic* response from your detector. Some extremely rich ores are in sulfides, tellurides, and other forms and, as these elements are *not* conductive, they will not produce a metallic response, regardless of richness. Most free milling ores, however, that contain non-ferrous metals in sufficient amounts and are conductive will produce a good response.

Generally speaking, about the only mineral that any type metal detector recognizes as "mineral" is magnetic black sand or magnetic iron ore ($Fe_3O_4$, magnetic iron oxide, $Fe_2O_3$, etc.). When the detector is tuned in the true TR discriminate mode, it is extremely easy to determine whether the ore contains a predominance either of metal or of mineral. If the specimen of ore contains *neither* metal nor mineral, your detector will produce *no* indication.

A "mineral" response from the detector does not necessarily mean that there is no metal present; rather, only that there is a *predominance* of mineral. If the specimen reads as "metal," you can be certain that it contains metal in conductive form in such a quantity that you should investigate the specimen thoroughly. These detector response capabilities make the low frequency TR discriminating metal detector (VLF/TR) the most important tool of today's successful prospector and miner for identification of metal *vs.* mineral.

## MAKE YOUR OWN METAL/MINERAL
## "ORE" SAMPLES

The first sample you will need is very easily obtainable. Acquire a silver dime for your positive sample. If you prefer, you may use a copper penny. The next two samples you will need will require the use of a little elbow grease.

The first is the mineral sample. Place a large iron nail or a piece of soft iron into a vise and file the nail with a coarse file, placing a piece of paper under the file to collect the filings. The quantity of filings you need is approximately equivalent to the quantity of filings you would obtain if a dime were completely filed into particles. Place the iron filings into a plastic medicine bottle with a diameter approximately that of a dime. Fill in with a white glue on top of the iron filings and let it dry. You now have a sample of non-conductive iron mineral that will cause a response identical to that caused by much of the iron mineral you will come across in your prospecting.

The next sample to make is one simply for the purpose of proving to yourself how extremely difficult it is to detect silver oxides, gold dust, and wire gold. Place a dime (or penny) in the vise and, with your file, reduce it completely to filings. Place those particles into a plastic bottle the same size as that in which you placed the iron filings, again running in glue to seal them permanently. You now have, basically, a marginal non-conductive, non-ferrous ore sample that will give a "questionable" positive response.

To assist your understanding of metal detector operation, learn the responses obtained from your samples. Such practice will greatly aid you in the field in analyzing detected veins and pockets and in learning how to identify the metal/mineral content of ore samples correctly. In addition to these three samples you should acquire, if possible, samples of the other ore specimens mentioned in this chapter.

Electronic prospectors high grade silver specimens at this old mine in Canada. They use the discrimination mode of the VLF/TR metal detector to high grade or select the conductive rock specimens. This ore sampling method can be successfully conducted with a quality VLF/TR type metal detector that has a precise, factory calibrated TR mode sampling point set at the "zero" discrimination point on the dial.

CHAPTER VIII
# How To Use the Latest Type VLF/TR Metal Detector for Electronic Prospecting

Only recently have recreational miners begun to understand the value of the new type VLF/TR metal detector for finding gold and other precious metals, in both conductive and non-conductive form. While it is impossible to guarantee instant success, it will be very difficult to fail if you observe three basic rules and are persistent in your search.

*First.* It is absolutely imperative that you choose the *proper type* of detector for prospecting. Here we are referring to the operating principle of the detector rather than to a particular brand.

*Second.* Patience is a must. Learn to understand your detector fully and become proficient in its use. Read the following pages carefully, practice with your detector, then re-read them. To be successful it is essential to understand *all* of the techniques described herein.

*Third.* Begin your search where gold is *known* to exist. It is impossible to find gold or other precious metals where they do not exist. Prospectors who have been successful with their metal detectors have employed wisdom and patience, plus a great deal of *research.* Stick to known, productive mining areas until you become familiar with your detector's operation and the telltale signs of mineral zones. Soon you will discover there are many things that can be done with a quality metal detector. It can open up completely new areas of success for the prospector and recreational miner.

## CHOOSING THE RIGHT DETECTOR FOR PROSPECTING

Without a doubt, the best choice of a metal detector for prospecting is a VLF/TR discriminating detector that has correct, factory-set, ground adjust and TR discriminate controls. This means the VLF/TR must be correctly factory calibrated for universal application for full metal *vs.* mineral capabilities. This type of detector is the deepest-seeking detector built. In addition, the VLF type of detector is a ground canceling detector. With a very simple adjustment, the operator can effectively cancel out iron mineralization present in the ground. This one factor alone permits operation in the heavily-mineralized areas where most mines are located. With the new VLF/TR detectors it is now possible to work areas that could not be touched with a metal detector only a few years ago.

No one has ever claimed that electronic prospecting was easy and required little effort for success. In fact, success usually comes only after much location research and long hours of field searching have been done. Here, Charles Garrett hunts the gold areas near Liberty, Washington.
Photo by Roy Lagal.

The VLF/TR ground canceling metal detector gives superior performance in the detection of small and large nuggets and both conductive and non-conductive ore (predominanently magnetic) veins. Such performance, along with the ability to cancel iron ground mineralization effectively makes the use of the VLF/TR ground canceling metal detector almost mandatory for prospecting.

While electronic prospecting may be done with a number of different types of detectors, you will have a distinct advantage if you use the VLF/TR ground canceling type. While the same general rules apply to several brands of detectors, you will have to experiment a bit to determine the precise response characteristics of the metal detector you own.

Correct identification of metal, mineral, and marginal ore samples is critical and has to be accomplished with the detector in a stable position. Otherwise, one is courting failure and incorrect sample identification. Detectors that discriminate or reject targets by operator manipulation (such as "whipping" or operating at a prescribed speed) have questionable use in the prospecting field. This is further attested to by the fact that under almost 100% of given circumstances the *sample* must be moved in a controlled manner to assure correct identification. It must be kept in mind, also, that operating quarters may be small and cramped so that "whipping" would be impractical. Basically, coin hunting type detectors that discriminate by "ignoring" the target or by operation in some unusual manner (such as "whipping") are of doubtful benefit in prospecting.

During the past several years much has been written about prospecting with metal detectors. The majority of that writing revolved around the tried, trusty workhorse of the industry, the beat frequency oscillator (BFO) metal detector. Until the advent of correctly designed and calibrated VLF/TR detectors, the BFO was the only instrument capable of performing all phases of prospecting correctly. The BFO has been responsible for finding many veins and many nuggets. However, it must now retire to second place in the electronic prospecting field. Those who used BFO instruments came to love and trust them but will now, with fond memories, delegate them to stand-by service.

## UNDERSTANDING YOUR SEARCHCOIL

Two types of searchcoils are commonly available on VLF/TR discriminating detectors: the co-planar loop and the co-axial loop. Both styles are available on the most advanced VLF/TR type detectors.

In large operations, bucket draglines rip down through the earth several feet below bedrock. Since most placer gold rests on bedrock, this type of mining operation is necessary. Dragline sites like this one are prime target areas for the electronic prospector equipped with the latest ground canceling metal detectors because it has been proved that oftentimes 100% of the gold is not recovered by these operations. It was at this site that Roy Lagal detected the arrowhead-shaped nugget.

The co-axial searchcoil is probably the most foolproof when it comes to electronic prospecting. This style of searchcoil gives very uniform response and it discriminates very well, but simply because of its shape and thickness it has never gained the popularity that the thinner, co-planar searchcoil has for coin hunting, nugget hunting, and other forms of treasure hunting. The co-axial searchcoil is a must in mines and other areas where 60-cycle and other electrical interference prevents normal detector operation when co-planar searchcoils are used.

## THE CO-PLANAR SEARCHCOIL

Because of the way the multiple windings are laid out in the co-planar searchcoil, it is extremely important that you locate and mark the exact electrical center of the searchcoil. Begin by laying your detector prone on a wooden bench with the searchcoil standing vertically at right angles to the stem. Remove all rings, watches, and other metallic objects from your hands and wrists.

Tune your detector in the VLF ground canceling mode to the threshold sound as described by the manufacturer's instruction manual. Then begin detuning the detector by rotating the tuning knob toward the "null" position until you get only a very weak signal from a dime passed close to the searchcoil. Now move this coin around on the bottom of the searchcoil until you get the loudest signal. If you have trouble determining exactly where this spot is, you probably still have the detector tuned too sensitively, and it should be detuned a bit more. With the tuning down low enough, it should be a simple matter to locate the area of loudest response. When you have located this area, mark a bull's eye on the bottom of your searchcoil with an indelible felt-tip marking pen. From then on, the bull's eye will be used any time you are identifying small ore samples.

## THE SECRETS OF SUCCESSFUL BENCH TESTING

With the bull's eye marked on the bottom of your co-planar searchcoil, you are well on your way to successful identification of metallic ore samples. However, there are two more little tricks to be mastered, particularly with *marginal* samples — that is, samples of low grade ore that contain only very small amounts of detectable metals.

First, *always* hold an odd-shaped sample in the same relationship to the searchcoil. An elongated sample (football-shape, for example) could give one particular reading if held broadside to the searchcoil and another if the pointed end is toward the searchcoil. This is true of *marginal* samples only and is due to the fact that the orientation may have more conductive surface lying horizontal to the search-

Roy Lagal, co-author, performs a bench test on possible silver specimen before transporting it home. Correctly conducted tests by the use of a TR circuit calibrated for ore identification will prevent many tons of worthless rocks being collected. This specimen was GOOD.

coil than when the sample is rotated ninety degrees. Flat metallic samples produce larger detector signals when the flat side lies horizontal with the searchcoil. On the other hand, iron mineral produces approximately the same negative signal, regardless of its orientation with respect to the searchcoil.

Second, always move the sample along the axis of the searchcoil *toward* and *away* from your center bull's eye marking on the bottom of the searchcoil. *Never* move the sample *across* the searchcoil . . . always *toward* and *away* from it. Reasonably quick motions produce speaker sounds that can be easily heard.

## CORRECT ADJUSTMENT OF YOUR VLF/TR FOR METAL/MINERAL ORE SAMPLING

There are two methods by which you can achieve metal/mineral ore sample identification. The most popular method is to use the detector in the TR discrimination mode; the other method is to use the detector in the VLF ground canceling mode. Both methods require that during production of the detector, the manufacturer makes correctly calibrated adjustments to the TR discrimination control and the VLF ground zero control. We will discuss the methods one at a time.

37

If the detector's TR discriminate mode is correctly calibrated at the factory, it is quite easy to achieve correct metal/mineral ore sampling. Rotate the TR discriminate control knob to its "zero" discrimination setting. Generally, this is fully counterclockwise. At this setting your detector should respond positive when *any* metallic object is brought up to the bottom of the searchcoil toward the bull's eye. Even tiny nails, regardless of their orientation, should produce a positive speaker sound and an upward deflection of the meter when they are brought toward the bull's eye. If the speaker sound decreases and/or the meter pointer decreases, or deflects downward, the discriminate control is not correctly set and you may not be able to achieve correct mineral/metal ore sampling. The reason that the control must be set as described is because, if it is set so that at its low discrimination level it rejects some metals and you then test low metallic content ore samples, the detector may reject them.

The second method (metal/mineral ore sampling in the VLF ground canceling mode) can be accomplished only if you can find the *correct null point*. The ground zero control must be correctly factory calibrated. By "null" we mean that, when the detector's tuning is set to "null," any metal,

A.M. VanFossen and Monty Moncrief trace and analyze a silver vein. There are countless thousands of active and abandoned gold and silver mines which contain veins and pockets of precious ore that was missed often by only a fraction of an inch, because miners could not see through solid walls. Modern electronic equipment will penetrate several feet through rock walls and most earth materials to locate conductive veins and deposits, as well as iron mineral veins that might easily contain rich ore that is not electronically detectable.

regardless of the kind, size, or shape, brought toward the bull's eye, causes the detector to respond with a positive sound and/or an upper deflection of the meter. When your mineral ore sample or any predominately non-conductive mineral ore sample is brought toward the bull's eye, the speaker sound should decrease and/or the meter pointer should deflect downward.

Those of you who have used the beat frequency oscillator detectors will know that the BFO has a fixed null at its center tuning point. This null correctly divides the detector's response between metal and mineral detecting. The null in a BFO never changes; it is always in the same place. The VLF/TR detectors do not have this fixed null. In order for a VLF to nullify correctly or "zero out" various densities of iron ground minerals, the null point must be adjustable. When you rotate the ground zero control, the null point changes its electronic position until the detector does not respond or produce a signal over the particular iron ground mineral where you are testing the detector. When this occurs, the null point is thus shifted from the "true" position and your ability to separate metal and mineral ore samples correctly is lost.

You can, however, quite easily find the "center" of tuning, or null point. Turn on the detector and set it to operate in the VLF mode. Bring your homemade mineral sample up toward (no closer than one inch) the bottom of the searchcoil toward the bull's eye. Rotate the ground control toward the metal setting until the homemade mineral sample causes the detector to respond positively. Then begin to rotate the ground control back toward the negative side. When you reach the point at which the ore sample barely begins to read negative, with absolutely no trace of a positive reading, you will have reached the true null of the VLF ground control adjustment. This, then, is the point at which you can achieve correct metal/mineral ore sampling in the VLF mode. This method, of course, necessitates that you always carry your mineral ore sample around with you or else remember the exact number of turns and from which "end" you must begin rotating the control to tune the detector to the true "null."

Generally, all ground zero controls utilize a ten-turn potentiometer. That is, the control knob must be rotated ten turns in order to achieve only one full rotation of the potentiometer itself. Notice as you rotate the control in either direction that it is difficult to determine exactly when you have reached the end of the potentiometer element. Carefully feel the knob's rotational friction. When you reach either end, you will notice a slight increase in

39

drag of the knob. This indicates the "end" of the control. So, then, when you need to check ore samples, *etc.*, you must locate the "null." As you have seen from the above instructions, you can carry your homemade mineral sample with you and find the null by using it, or you can remember how many turns, for each searchcoil, you must rotate the ground control from your chosen ground control "end." One way to keep track is to write the number of turns on each searchcoil.

To sum up, if your detector is correctly calibrated at the factory, you can check ore samples accurately in the TR discriminate mode by rotating the TR discriminate control to its lowest discrimination point, that being the point where there is no discrimination. If your detector is not equipped with the TR discriminate mode, or if you wish to check ore samples in the VLF ground cancel mode, you must find the center of tuning, using either of the two methods described above — rotate the ground control the correct number of turns from one "end" of the control or use your homemade ore sample to find the null.

## ANALYZING YOUR SAMPLES

The best way to understand how your detector is reacting to different ore samples in the field is to do some experimenting on the test bench. Use your homemade samples and also obtain some samples of galena (lead), silver and gold ore, iron pyrite (Fool's Gold), and ordinary rocks. If possible, get two grades of ore for your tests, both a very high grade and a very low grade specimen of each of the above materials. By conducting your own bench analysis you will become familiar with the type and amount of detector response to the various materials.

If the ore you are checking has a predominance of metal in detectable form (you can use your solid dime sample for testing), you will hear a sound increase as the sample is moved closer to the searchcoil. If your detector is equipped with a sensitivity meter, you will see a positive pointer movement indicating the presence of metal. On the other hand, if the sound has a tendency to "die" slightly, the ore (or your iron filing sample) contains a predominance of mineral or natural magnetic iron ($Fe_3O_4$). As mentioned previously, this does not mean that the sample contains NO metal, only that it contains MORE mineral than it does metal. If the sample contains *neither* metal nor mineral, or if it has electrically *equal* amounts of both, you will receive *no* response on the detector.

Remember, metal detectors do NOT respond to many types of ores. Only those containing metal in conductive form in sufficient quantity to disturb the electromagnetic

40

Jack Kirsch (right) watches as Charles Garrett, owner of Garrett Electronics and author, carefully evaluates the response received by a ground canceling, deepseeking type detection instrument. Roy Lagal, author of many books and without question the most knowledgeable of the electronic prospectors, is trying to determine the predominant content of a vein or dyke he has just detected. Charles travels extensively throughout the world and takes great pains to check out new detection instruments and methods. The new ground canceling detectors and Charles' and Roy's sharing of the knowledge they have acquired about the capabilities of the equipment have helped create world-wide interest in gold hunting with detectors.

field will cause the detector to respond positively. This initial bench testing will help you become familiar with the type and amount of response your detector gives to both low grade and high grade ores. A little time spent experimenting now will save you many hours of confusion when you get in the field!

## WHAT ARE "HOT ROCKS"?

As previously discussed, you often will detect hot rocks. When the detector is in the VLF ground canceling mode, the signal you receive will be the same as you would receive if you had just detected a metallic object. The signal is positive and unmistakably "metal." As with any detected

41

targets, you must properly identify the target before spending time digging. The method to use to identify hot rocks is explained in detail in Chapter X. We mention it in this chapter in order to bring you up-to-date on the nature and characteristics of hot rocks. Hot rocks are found in jumbled rock piles, as well as in ground areas that look as if they have never been disturbed. A hot rock is a rock that contains a quantity and/or density of non-conductive iron mineral such that it causes a "metal" response when the VLF ground canceling control of your metal detector is adjusted to some given point with relation to the detection characteristics of that rock. In other words, as you use your metal detector in the VLF mode over mineralized ground, you adjust the ground control to a certain point so as to nullify the detection capabilities of the detector to the earth minerals over which you are operating. However, when a sufficiently different (excessive content) iron mineral hot rock is present in the soil, it can cause the detector to respond to it positively.

This is further explained by saying that no hot rock will produce a false signal on a BFO because the BFO has a fixed null or center of tuning. This null never changes, as we have noted; therefore, any non-conductive iron ore always gives a negative reading on a BFO. However, as you adjust the VLF ground control, the null or center of tuning changes its position, depending upon the quantity and/or density of the iron earth minerals. As you shift the VLF mode tuning beyond the "BFO" null into the *metal* detection zone of the VLF detector, then certain mineralized rocks (hot rocks) will respond as metal. These detector characteristics may be troublesome to you at first, but after you have adjusted your detector and correctly interpreted a few responses to hot rocks, you will be able to identify them quickly with the TR discriminate mode and you will shortly feel right at home with your detector.

## GRID PATTERN SEARCHING

Grid pattern searching is a successful method often used in field prospecting when looking for veins. Sweep the searchcoil in wide, even strokes according to a fixed grid or crisscross pattern. It is an excellent technique since it gives you the opportunity of approaching a vein or deposit from two different directions, avoiding the distinct possibility of walking parallel to a vein and never actually crossing it.

Adjust your detector (VLF mode) to the surrounding ground matrix level of mineralization. (DO NOT USE AUTOMATIC TUNING.) Walk slowly, in a straight line

42

if possible, scanning a wide path with the searchcoil ahead of you. When you get to the end of a "line," turn and walk a parallel path approximately ten feet from the first path back in the opposite direction. Continue until you have covered the area. Repeat this procedure, except walk parallel paths at 90-degrees to the first set of paths. Now you have completed your fixed grid or crisscross pattern search of the area.

If your detector speaker sound increases and/or the meter pointer increases however slightly while you are scanning, notice if the increase in response remains at the increased level. If it does, you may have a vein or ore pocket or the detector may have changed its tuning due to temperature, bumping the controls, *etc.* To check it out . . . DO NOT CHANGE THE TUNING and go back to a point just before you noticed the change. If the speaker and/or meter DECREASES TO THE PREVIOUS LEVEL, the detector response change was not due to some detector or operator problem.

Roy Lagal watches Charles Garrett pan a sample taken from a hot spot reading that Roy obtained with his detector. Often, bullets and iron pieces are located. Lead bullets cause a reaction the same as gold, silver, and copper. Iron targets can be analyzed with the TR mode and ignored. When targets are detected in the VLF mode, a quick flip to the TR discriminate mode quickly identifies the target as either a good or a poor conducting (iron) target or as one containing a predominance of black sand which should be panned.

Retrace your steps. As you reach the point where the sound changed the first time, it should again change if you are detecting an ore vein or pocket. As you continue walking, pay close attention to the detector's audio and meter response. You should, at some point, notice that the detector sound either increases further or drops off to your initial tuning level. If the detector response increases, the ore is getting "richer." When the detector response falls off to your initial tuning level, you have walked on over or past the ore deposit. Plot or map the deposits or veins (if there are several) as they cross and crisscross each other beneath the surface.

After you have found detectable ore deposits, you can identify their nature. To identify the deposits, you must operate the detector in the TR discriminate mode. The TR discriminate mode will identify whether the deposit is predominantly iron (non-conductive pyrites) or predominantly non-ferrous material. Veins crisscross each other beneath the surface . . . a gold vein may be bisected by several iron veins. You can, fairly accurately, plot these veins by paying careful attention to the responses of your detector. You will also receive positive metallic indications from heavy concentrates (pockets) of magnetic black sand. Often, these pockets will contain gold and, of course, should be investigated. See also VEIN AND ORE DEPOSIT MAPPING, Chapter XI.

For a detailed study of various other detector types and their applications in prospecting, obtain a copy of DETECTOR OWNER'S FIELD MANUAL, by Roy Lagal, from Ram Publishing Company, P. O. Box 38464, Dallas, Texas 75238. This book fully explains the characteristics of the BFO, TR, the PRG, and the PI detectors and their applications.

# CHAPTER IX

# Successful Nugget Hunting Techniques

## NUGGET SEARCHING IN WATER

While some manufacturers provide smaller searchcoils for nugget hunting, careful field testing will show that 7- to 8-inch VLF/TR searchcoils are very efficient, *even for locating the smaller nuggets.*

Following the manufacturer's instructions and carefully adjust the ground canceling control of your VLF detector to eliminate interference from the highly mineralized black sands found in most gold-bearing stream beds. Use earphones to heighten your awareness of even the faintest detector signals. Operate the searchcoil from four inches to one foot above the bottom of the stream, moving it slowly over the search area. The operating height will depend upon the mineralized content. Make certain your detector has *submersible* searchcoils.

Gold and silver nuggets, pockets, and veins are not always found in remote, out-of-the-way places. Oftentimes we walk right past nuggets lying just a fraction of an inch beneath the surface, as was this huge nugget that weighed 36 ounces and is valued at more than $50,000! This beautiful gold nugget was finally removed from the resting place where it had lain for eons of time.

Two tools will be of great assistance for recovering gold nuggets from the water: (1) a plastic gold pan and (2) a small garden trowel or shovel, depending upon the depth of the water. When you get a metallic target reading, carefully slip the shovel under the coil, lifting the sand and gravel contents into the plastic pan. Test the entire pan of material with your detector, as described earlier, to see if you have indeed recovered your target. If the target is not in the pan, dump the material back into the water, locate the target again with the detector, and scoop another shovel of gravel into the pan to check with the detector. Continue this process until you find the target.

When the metallic target is in the pan, try visual recovery first by sorting the gravel carefully. If you cannot locate your object in this manner, concentrate the material by panning. If you are still unable to spot the target, perhaps it is only a small piece of ferrous trash. There is also the possibility it could be a highly mineralized hot rock with more $Fe_3O_4$ content than that for which the detector was adjusted. To clear up the mystery quickly, electronically check out the target according to the procedures given earlier.

This same method of detection may also be used on old dredge tailings either in the water or on the banks. Dredge tailings are easy to spot due to the uniform size of the gravel piles. Old dredge tailings have produced some fantastic finds for treasure hunters using metal detectors so don't overlook this golden opportunity!

## NUGGET HUNTING IN DRY PLACER
## DIGGIN'S AND DRY WASHES

Old placer diggin's and the bottoms of dry washes are often productive and rewarding locations for nuggets. In remote desert areas where water has never been available and where the only method of recovery is with a dry washer or by dry panning, there are millions of dollars worth of small nuggets. These nuggets are rarely detectable by eyesight, but they lie almost on the surface or at very shallow depths. Investigation of low-lying areas with a highly sensitive, ground canceling metal detector can be very rewarding. Some knowledgeable nugget hunters have done very well over the years by working dry or desert areas in highly mineralized locations and by working areas where large gold nuggets were found by early day prospectors. Do your research! The use of a metal detector is practically the only method of locating nuggets in these areas; it is certainly the quickest.

Again, the VLF/TR type detector is the best choice for nugget hunting. The 7- to 8-inch coil will give the

Roy Lagal and Virgil Hutton explore some old placer diggings. These small tunnels drifted in on bedrock following a fault or crack by the early-day Chinese miners. These are prime target areas for the electronic prospector with a good metal detector. However, there are literally millions of tons of gravel concentrated above the tunnels, and sometimes the least noise will cause a cave-in. Bones of early-day miners are often found in such rediscovered tunnels. *Don't leave yours there.* If you use caution and a good partner, it is possible to search such dangerous places profitably. A detector will scan a few inches deep on each side of the tunnel and often produce more by this method than was recovered by the one who expended all the labor to drift the tunnel.

greatest depth, yet is sensitive even to small nuggets. The ground mineralization canceling circuitry will certainly make this type of detector the easiest to use.

After you have correctly tuned out the ground mineralization according to the manufacturer's suggestions, you should scan with the searchcoil held an inch or two above ground surface. CAUTION: The searchcoil operating height will always be determined by the amount of mineralization present. Greater than two-inch operating heights may be required. Be confident that your VLF type detector is doing a good job for you. Be aware that all detectors are *not* suitable for prospecting and that poor results yielded by some detectors have actually caused the hobbyist to quit this interesting and profitable pursuit. The VLF circuitry will penetrate black sand with ease and pick out the nuggets.

Again, you may run into trouble with highly mineralized, out-of-place hot rocks and, again, the TR discriminator circuit of your VLF will enable you to unmask these cleverly concealed little nuisances instantly. (See Chapter X.)

47

The recovery of nuggets from dry sand is achieved basically the same as from a stream. Slip your shovel under the sand where you get the signal; place the material in a plastic gold pan and check it out with your metal detector. If you fail to get a reading, dump the pan and repeat the process. When your detector verifies the target is in the pan, make a visual search. If you are unable to locate the target, dry pan until you find it.

A good VLF/TR discriminator will amaze you with the number of things it will enable you to recover from the desert. A little practice and experience will make you an expert in no time!

Even though prospecting methods in certain parts of Mexico remain the same as those used centuries ago, Mexican prospectors can still earn a fair living. Charles Garrett points to a silver nugget this Mexican found. Note the wooden gold pan. The can lid Charles is holding and the small, round metal pan contain several ounces of silver and gold found during a few hours' panning in a near by river. Charles' parting gift to the man was a "Gravity Trap" pan. In return the man handed Charles a silver nugget which is displayed in the Garrett museum in Garland, Texas.

# CHAPTER X

# *How To High Grade a Mine*

OLD MINES CAN STILL YIELD RICHES

If you find a likely-looking mine, begin by checking the floor with your metal detector. There is a possibility of finding high grade ore in older mine tunnels and shafts because they were worked by less modern methods than the new mines.

Although very few recreational miners seem to realize it, the tunnel floors of old mines can be some of the most productive areas for making a metal detector search. *All* the ore ever taken from the mine had to cross the tunnel floor. It is almost certain some of it fell off the ore carts and was eventually covered with rock and debris. This high grade ore has laid there unnoticed for years, just waiting for some lucky treasure hunter to find it.

Begin your search by tuning the ground canceling circuitry of your metal detector to cancel the ground mineralization. Make a general sweep of the tunnel floor, collecting all samples that give a metal reading. Remember, there are

After a few minutes of chipping at the wall, the miner exposes and breaks loose this silver pocket Charles Garrett detected. Charles still remembers that when the detector searchcoil passed over this spot, the speaker sound, reverberating throughout the mine tunnel, hurt his ears. This small, isolated pocket was the only detectable silver along the wall. Think how many such mining tunnel walls are still unprospected by electronic means!

likely to be a lot of small iron objects in the old mine tunnel — rail spikes, pieces of the old rail itself, and hangers for candles or mining lamps. Spikes were also used to secure the tunnel shoring, and these will be picked up by your metal detector.

Once you have collected a number of likely-looking ore samples, place your metal detector in the prone position and tune it as you did in your bench testing. Individually test all rock samples. Do not limit your testing to the samples you found during the initial sweep of the mine. Work the tunnel thoroughly from one end to the other. Use your rock pick to dig beneath the top debris and test all small, likely looking samples with your detector.

Here is an additional tip on this type of mine searching. The "pay streak" will normally always contain a much higher metal or mineral content than the ordinary rock of the tunnel. For this reason, save *any* unusual samples for later examination.

## FINDING MINE POCKETS AND VEINS

Many times a mining operation would miss a pocket or vein of high grade ore by mere inches, chewing on into the mountainside in a sightless search for new ore deposits.

Electronic prospectors scan a wall where a silver pocket had been taken out. It was hoped additional silver that the miners had missed might be found. Don Garrett (author's brother) walks up with silver sample he found on the floor.

It is well within the limits of modern VLF/TR discriminator type metal detectors to pinpoint these valuable pockets and veins for the electronic prospector.

If the mine tunnel has been driven through highly magnetic mineralized material, the ground control of the VLF detector must be adjusted to cancel the effects of this mineral. Then tune the detector so you can maintain a constant but faint sound from the speaker. Operate the searchcoil approximately four to twelve inches from the tunnel wall, depending upon the amount of iron mineralization present. Scan the walls and ceiling carefully, marking or taking note of any positive (metallic) signals. (A can of spray paint can be used to advantage to make the markings.) Ore containing a sufficient amount of conductivity (and some non-conductive magnetic ore) will respond as positive (metal).

## HOW TO IDENTIFY "HOT ROCKS" AND "HOT SPOTS"

Tuning the ground control to compensate for a high mineral background will cause the detector to respond positive even more often to hot spots (and hot rocks) that are isolated or out-of-place with respect to the mineralized area for which the tuning was originally set. (See WHAT ARE "HOT ROCKS," Chapter VIII.) The identification of hot spots would seem to present an insurmountable problem; however, because your searchcoil sees these hot spots at shallow depths only, the TR discriminate circuit of your VLF detector will immediately expose these bogus readings by rejecting the hot spots (provided the detector has been correctly calibrated to permit universal use, including prospecting).

The procedure for identifying mineral hot spots (and hot rocks) is a simple one, but it does require practice. To check to see if the "metal" response may be a mineral hot spot, pinpoint the target using the VLF mode. The move the searchcoil to one side, lower the searchcoil slightly or set it on the ground, and switch to the TR mode of operation. Push the button to retune. (Flip-switch Master Controls retune automatically). With a constant sound coming from the speaker, pass the searchcoil back over the target, keeping the searchcoil at exactly the same distance from the ground or wall as when the hot spot was first located. You may have some difficulty at first when attempting to keep the searchcoil at this constant distance to achieve a steady audio response, but a small amount of practice will overcome this problem. If the sound decreases or stops suddenly, the target is magnetic iron ore or oxides. This is the *only* substance that will cause the signal to stop. If this happens, ignore the target, switch back into VLF, and

51

continue your search.

If, on the other hand, the signal increases, remains steady, and/or shows no inclination to decrease, the target warrants further investigation. The *variable* discrimination control is now increased to determine the amount of conductivity of the target. (Note: It is best *not* to use detectors with fixed or discrete TR discriminate controls.) If you have previously practiced with your TR discrimination control, you already know the approximate point on the control where worthless PYRITE will be rejected. If you still receive a positive response after you have passed beyond this setting, it is very possible that you have discovered a non-ferrous pocket or a vein of conductive ore.

Remember, rich iron pyrites have a much *lower* (poorer) conductivity factor than non-ferrous ore, and the pyrites can be rejected by the VLF discriminator circuit while it will still accept low grade non-ferrous ores. Also keep in mind that the ore body may *not* be solid metal, and the response may be rather faint unless the ore is of extremely high conductivity. (Your homemade non-ferrous ore sample proves how difficult it is to detect non-ferrous, non-conductive ore.) Don't overlook those weak signals.

VLF/TR ground canceling detectors act very similar to anomaly detecting instruments in that they will detect, with "metallic," positive indications, all shallow veins, pockets, stringers, *etc.*, where sufficient quantities of magnetic iron are present. When the VLF/TR detectors are operating in the VLF ground canceling mode, with the ground control advanced at least twenty percent (usually one turn past null into the positive region), then these veins, pockets, stringers, *etc.*, will produce positive reactions from the detector. This important detection capability is somewhat akin to anomaly detection. In other words, even precious metals in a non-conductive form (tellurides, oxides, etc.) present in magnetic veins can be detected using this method. The magnetic vein need only contain enough ferrous iron for detection purposes. Of course, the ore content must still be determined by analysis or assay, but this important capability of the VLF/TR is largely overlooked even by professional geologists.

Many of you who have utilized BFO detectors in the past should consider returning to areas you worked previously, especially if you walked on past "negative" BFO readings, thinking you were detecting magnetic iron only. Rich gold, silver, and veins of other substances often contain large percentages of magnetic iron. Of course, in areas where there are no known precious metals or minerals, in all likelihood these signals can be considered to be magnetic iron only. However, in areas where it is likely that

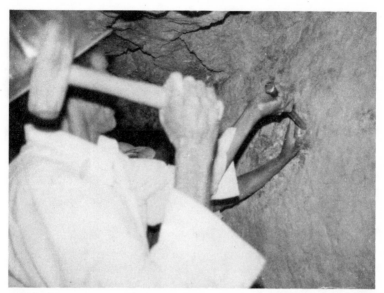

The miner in the top photograph begins to chisel an opening where he will place a stick of dynamite. Charles Garrett's detector has responded with a signal indicating a conductive, non-ferrous substance. In the lower photograph, the miner points to a one-half-inch-wide native silver vein which the dynamite charge exposed. The vein continued downward through the mine wall where it eventually became several inches wide. This vein might still have been undiscovered a thousand years from now if it had not been located with an electronic detector.

Javier Castellanos displays a few slivers of beautiful native silver he recovered from a vein in one of his silver mines. These mines, located in the state of Chihuahua, are fabulously rich. Countless veins and pockets await the modern-day prospector equipped with the proper type metal detector. Truly successful electronic prospecting has been achieved only since the introduction of ground canceling detectors. We are now on the verge of a world-wide gold rush that might make that of the last century look like a picnic!

rich veins are located, then blasting, core drilling, *etc.*, could be warranted. Those serious about this aspect of prospecting should consider obtaining core drill and assay kits.

Elsewhere in ELECTRONIC PROSPECTING we describe the method by which you can analyze detected substances by utilizing the TR mode. Keep in mind that the operation in the TR mode can be unerring in the identification of conductive metals, but beyond that nothing can take the place of an assay.

### A WORD OF CAUTION

When you are exploring or working around deserted mine shafts and tunnels, extreme caution should always be exercised. Abandoned mines can be very dangerous. The shoring timbers have rotted over the years and earth movement and water seepage may have loosened once-solid tunnel walls. Detector sounds or any loud sound or impact against the timbers or walls of a tunnel could bring the mountain down on top of you.

Equal care should be used any time you lean over to peer down a mine shaft. Not only is there a chance the earth could slide, tumbling you hundreds of feet straight down into the bowels of the earth, but poisonous fumes coming up from shafts have actually killed people instantly — people who only wanted to get a look at an old shaft.

Never work alone. Never let all the members of your party enter a mine at the same time. Someone should always be stationed at the entrance to summon help should it be needed.

Also keep in mind that you do not necessarily have the right to enter and begin searching just because an old mine *looks* deserted or there is no one around. There are a lot of old-timers with valid claims who just might object to your doing a little high grading on their property. Sometimes those old-timers object a little violently! It is always best and proper to gain permission to search. Very few claim holders will object to your doing a little electronic prospecting, particularly if you offer to give them helpful information on any veins or mineral pockets you might locate.

## SEARCHING OLD MINE DUMPS

As you travel through the deserted gold fields you will see many abandoned tunnels, rock heaps (called "tailings"), and piles of ore that never made it to the stamp mill to be crushed. By using your metal detector properly to thoroughly test these rocks, one at a time, you can often recover many good ore samples.

Charles Garrett, co-author, and Virgil Hutton find a large nugget in some old tailing piles. The previous two days produced nothing until they worked their way almost to the north side, or end of the tailings. Often acres will be barren, and only a small portion of the dump will contain any value. This is often caused by the previous dredge operator's lack of attention to large nugget recovery, or their lack of knowledge that nuggets were present. Regardless, if research proves that you are in an area where large nuggets have been found by any previous mining method, it definitely pays to use a VLF/TR metal detector on such tailings. The manual TR circuit is used to identify "hot rocks" versus metallic targets, whereas the VLF circuit is used to conduct your initial search. Remember, it takes only one good nugget to make a week's pay.

One important thing to remember when searching an old mine dump that may contain high grade ore is that during the working period of the mine perhaps there was only a certain small portion of the dump that actually received tailings from the vein itself. The rest of the dump may be only debris from the mine shafts and tunnels. Thus, you can readily see it is important to take rock or ore samples from many different locations, especially from the higher sections of the dump.

To check the dump for high grade ore samples, lay your detector prone and tune it just as you did in your bench testing. Then move your selected samples, one at a time, quickly *toward and away* from the coil to test for metal. After a reasonable number of samples from one location have been tested and no metal has been located, move to another area of the dump. If, after prolonged testing, you do not recover at least a few metallic specimens, it may be that the ore may have been of a very low grade composition and is of little value.

Do not overlook the possibilities of working these old ore dumps. Many beautiful silver and gold specimens (of considerable value, we might add) have been recovered this way, hidden inside of worthless-looking chunks of mine tailing debris.

CHAPTER XI

# Field Prospecting

"Field prospecting" encompasses a wide variety of searching for precious metal, such as locating deep veins, looking for placer or nugget deposits, hunting for rich float material, and pocket hunting. Field prospecting is where all the knowledge you gained previously in your bench tests will pay off in big dividends.

## FIELD SEARCHING FOR ORE VEINS

Contrary to popular belief, large mining corporations do not do nearly the amount of field prospecting most people think they do. This is due mainly for the reason that these large companies are highly profit oriented and, since prospecting takes a great deal of time, they rely on individual prospectors to make their finds for them. If you should be fortunate enough to locate a rich ore vein or deposit, it can oftentimes be sold to one of these corporations for a great deal of money. The recreational miner, who enjoys just being in the wilderness for a relaxing weekend whether or not he finds gold, is more inclined to hunt for ore deposits than are the large companies. The search for gold will take the weekend prospector into some of the most beautiful areas in the country, offer him the chance to camp in fantastic natural settings, and still present the opportunity to come up with a really "big one."

Since all types of gold deposits originate as an ore vein formed during the volcanic activity of past eons, such veins may go deep into a mountain and be fabulously rich. When such a vein is located, mining companies will literally jump at the opportunity to purchase this type of claim, extract the ore from the vein, and smelt it down into pure metal.

## DEEP VEINS

Deep veins are usually a composition of several metals and minerals. For this reason, extreme care must be exercised because the signal from your metal detector may be very faint.

A vein may be either metallic and respond as metal or it may have a predominance of iron oxides. Regardless, your detector will respond positively to it. Evaluate the signal by considering its magnitude, in which direction and how far it runs.

57

The use of the largest coil with which the detector is equipped is absolutely necessary since veins can run very deep. Pay close attention to all responses. Investigation of irregular or unusual signals will many times lead you to pay dirt.

## LOCATING <u>BOTH</u> MAGNETIC AND NON-MAGNETIC VEINS, POCKETS, *ETC.*

A tremendous VLF/TR detector advantage when operating in the VLF mode is the detector's ability to detect both magnetic (ferrous) and non-magnetic (non-ferrous) veins or pockets simultaneously. BFO and TR types indicate only the approximate *difference* in response between magnetic (ferrous) and non-ferrous content. This presents an almost insurmountable problem due to the fact that nearly all veins and pockets in highly mineralized areas contain either a large or a predominant amount of ferrous iron. The actual precious metals (non-ferrous) in most veins comprise perhaps only one-hundredth of one percent, yet the veins are considered extremely rich. This fact has caused the standard metal detector to be used only sparingly over the preceding years in mining exploration. The introduction of the VLF type for public use has changed the picture to one of high optimism for the more modern prospector.

A miner chisels a sample loose at a spot where a strong positive reading was obtained with a VLF/TR detector. This spot continued to indicate the presence of a highly conductive metal, even though the TR discriminate mode was adjusted far into the reject section of the TR discriminate mode. Readings like this mean money in the bank!

When tuned to the surrounding matrix in a highly mineralized area, the ground canceling VLF types will indicate the presence of EITHER OR BOTH veins or pockets containing conductive metals or a higher content of ferrous (non-conductive) content than that to which the ground canceling circuit was previously adjusted. Prior to the introduction of the VLF detector, practically ALL ferrous indications have been ignored as worthless. With this increased knowledge of the VLF type, perhaps many of the so-called "iron" indications will be investigated and many rich discoveries will result.

It is possible to narrow a search area somewhat by first finding surface float. Use small coils when searching for float. When you think you have found the vein's location, change to the larger coils and work in a grid pattern. Set the detector's tuning so that there is a slightly audible (threshold) sound and make wide sweeps with your searchcoil. The response area may be quite wide and, if you fail to cover enough area in your sweep, you may not be able to determine the edge (start and finish) of the signal. Within reason you should be able to judge the depth of the vein by the amount of response from your detector.

## VEIN AND ORE DEPOSIT MAPPING

Once you have detected an ore deposit you will find it quite interesting and intriguing to map or plot your discovery. You may have discovered an isolated pocket several feet in diameter or a very intricate network of crossing and crisscrossing iron and non-ferrous veins.

Follow the grid pattern searching techniques described in Chapter VIII, paying careful attention to detector responses as you plot them on paper. (Remember, do not use automatic tuning.) You may find the response increases slightly and then remains at a constant level for a distance and then returns to the preset tuning level. On the other hand, you may find the detector changes its response quite often, indicating, possibly, a network of veins. The vein responses can be narrow or wide, depending upon their nature and size and/or how the iron oxides have leached out into the surrounding earth matrix. A vein one foot wide can leach out such that it appears several feet wide to the detector.

Carefully evaluate the areas over which the detector response changes several times. While standing over the vein or deposit, retune the detector to your preferred operating audio level. Pay close attention to the detector's response levels as you walk parallel with and on top of the vein or deposit. Try to determine if the response changes and how much. Oftentimes certain changes will indicate several magnetic ore pockets or perhaps contact points where veins bisect an ore deposit or other vein. In areas that have already been worked by prospectors careful readings can many times tell an experienced miner of an extremely rich pocket or a place where a rich vein has "slipped" and become "lost" from known pockets or veins. The seasoned prospector will always correctly identify all deposits. Core drilling is often the easiest and fastest way to obtain samples for assaying.

## PLACER DEPOSITS

Erosion by wind, water, landslides, *etc.*, are continually breaking down the rock deposits in gold country and releasing gold from its hard rock prison. Run-offs from melting snows or heavy rains move the gold along as the water rushes down mountain slopes into small streams, into larger streams, and finally into rivers. During these periods of high volume water movement, the streams and rivers in gold country are heavy with dirt, rock, other debris, *and gold!* As the water rushes along, the heavier materials, including the gold, become trapped along the way, while the lighter dirt and rock are swept on downstream.

Authors Charles Garrett and Roy Lagal investigate a bend in the river at the Canyon de Cobre in Old Mexico. Located in a rich silver district, the authors discovered a small "high bar" that the Spanish conquistadors missed. They spot panned the area and recovered about twenty ounces of pure native silver. In long crystal form it is beautiful to behold. Time forced departure, and they showed the local Indian prospectors the placer location and gave them some plastic gold pans. They also instructed the Indians in the use of the "Gravity Trap" riffles and were rewarded by a gift of a primitive wooden gold pan used by the isolated tribe, and also the possible location of some more old Spanish diggings. Going back? You bet.

Nature has provided literally thousands of places along the way where placer gold can become trapped. Due to its weight, gold is pulled by gravity into pockets (hollow depressions) and cracks and crevices in the stream bedrock. It is often entangled in tree and grass roots and deposited behind huge boulders.

Placer mining in rivers and streams is easier than dry placer or hard rock mining because the fast moving water has already done part of the work for you by sifting and sorting the rocks, gravel, and gold. The successful placer miner will soon learn to spot these traps and check them out with his metal detector and gold pan.

Placer mining is probably the most popular type of prospecting for the recreational miner because his chances for success are better, less equipment is required, and the exercise is healthful and enjoyable. On the other hand, hard rock mining of ore veins requires large amounts of equipment, more work, and the movement of tons of surrounding rock to recover the gold. This is why most recreational miners sell their hard rock claims to large companies and spend their time at the more enjoyable project of placer mining, especially with the advantages of electronic prospecting that are now possible.

## DRY WASHES ALSO CONTAIN GOLD

After years of motion picture and television stories, a lot of people think gold is found only in streams and rivers in the gold bearing states. This is far from true. Granted, it is easier to recover gold if there is a good supply of running water, but water is not an absolute necessity. In the old days, placer miners used water to pan the sand and gravel from the heavier gold, but with today's modern gold pan with its built-in gravity traps panning can be productive even without water. The old-timers almost always followed the stream beds because it was there that the water carried the gold down from the mountains.

What about the old dry washes that had not seen water since a stream changed its course scores of years earlier? These dry beds also contain gold, but until the development of modern metal detectors the idea of having to find the gold by panning acre after acre of dry ground was a most discouraging proposition. With today's ground canceling metal detector prospecting is much easier, and the dry washes which were passed up in the 1800's can prove very productive.

Also productive are the old placer diggin's. An early-day placer miner could have been standing right on top of a large nugget and never have spotted it since he had only

61

his eyesight to use as a locating tool. A recreational miner today has the ability to locate nuggets even though they are buried under a layer of gravel or sand. By using the detector to scan the old diggin's and placer mining areas, gold that has been passed by for centuries can easily be located. In most instances, the use of a correctly designed, quality metal detector of the appropriate type is the only method by which these long-overlooked deposits can be located. In *all* instances, the use of a detector is the fastest, most practical way to pursue recreational mining. With a quality detector, one designed for universal use including prospecting, you can cover hundreds of times the ground that you could with a shovel and pan.

## LOCATING MAGNETIC BLACK SAND POCKETS

Black sand concentrations do not necessarily contain gold or other metals of high specific gravity, but if either or all are present they do tend to settle and become trapped in natural pockets. Operators of small surface type suction dredges have used the BFO type detector for years to locate these concentrations. The BFO was marginal for this use in highly mineralized zones due to the erratic response caused by variation in searchcoil height over the uneven search area. Since most streams and dry washes are generally unlevel to some degree, and most search areas are located in highly mineralized districts containing a high content of magnetic minerals, the VLF ground canceling type will operate easily over the unlevel surface, locate the heavier black sand concentrations, and at the same time detect any large nuggets present. (This would be impossible with any type other than a ground canceling VLF circuit.) Regardless of the amount of magnetic mineral present, follow these procedures: Operate in VLF mode; adjust background mineralization out same as normal procedure; now, advance ground control at least one or more full turns into the positive side causing the ground or "gravel" to now respond as metallic. Carefully maintain an even operating height, permitting the detector's audio to respond to targets. (When the sound is allowed to respond full force, the detector cannot respond further when targets are encountered.) More operating searchcoil height must be allowed in heavy mineral areas: generally, four inches to approximately one foot. The detector will now detect large pockets of magnetic black sand and will also respond to any detectable metals, gold nuggets, or trash metal. Of course, "hot rocks" are detected at all times, and these must be dealt with the same as any other undesirable target.

Practice this procedure before attempting field operation: fill a cloth bag with approximately one gallon magnetic black sand. Operation will be the same in all areas. Adjust VLF mode to null out mineral content of background. Place bag (containing sand) in search area; place large nugget or metal object in search area;

place "hot rock" also and scan the targets. The metal targets (gold, trash etc.) and the "hot rock" will respond where the black sand will not. NOW, advance the ground canceling control one or more full turns into the positive side until the ground responds as "metal." Maintain an even operating height causing the detector's audio to remain fairly even and re-scan the targets. All the targets now respond as metal, including the bag of black sand. This gives a tremendous advantage to the operator who takes the time to discover how simple, and easy it is to locate mineral pockets by their magnetic content, AND, not miss nuggets during the operation. This easy procedure can be accomplished either in wet or dry areas.

## FLOAT AND POCKET HUNTING

Float is ore that has broken off from the mother lode and been carried down hill by gravity, wind, water, earth tremors, and other acts of nature until it has lodged in hollow places, behind boulders and other barriers. These heavy pieces of ore form pockets on hillsides and are often covered by an overburden of rock and gravel that has washed in on top of the gold.

A good VLF/TR detector is the best equipment for locating rich pockets of metallic ore. Medium to large searchcoils are preferable because you will need the added depth capability, as well as the larger coverage of the ground to speed up your search.

Begin your search for pockets or "float" on depressions in the hillsides or in gullies, creek bottoms, and in other areas where you believe the heavy float had to stop. Logical reasoning will dictate where to search. Once you begin to locate float ore, work upstream or uphill from your initial discovery in an attempt to locate the source. Remember the ore is a combination material and will not produce as strong a reading as will a solid, metallic object. Do not overlook any faint signals. When you get a reading over a jumble of rocks, lay the detector prone and proceed to move the rocks, *one at a time*, quickly toward and away from the coil as you practiced in your bench tests. This procedure will enable you to tell exactly which rock(s) produced the signal.

When working a creek bed or hillside, adjust the detector to cancel any ground mineralization present and sweep the searchcoil two to six inches above the surface, depending upon the amount of iron mineralization present. Listen carefully for faint indications. The signal may be faint because the pocket may be covered by a thick layer of overburden *and* because the ore may *not* be pure metal. (Some iron mineral may also be present.) Nevertheless, if the pocket is relatively rich (conductive) you will most

likely get a positive, metallic response from the VLF
detector.

# CHAPTER XII

# The Rockhound and the Metal Detector

In the hands of the rockhound, a good metal detector can be an invaluable aid, not only for helping to identify certain specimens but also in locating valuable rock samples. Properly used, the detector can prove very rewarding, but it should not be considered the ultimate answer to the positive identification of all minerals and gems. Here nothing can replace the knowledge gained from experience in identifying semi-precious stones and gems.

The metal detector is a valuable piece of the rockhound's field equipment and is capable of aiding in the locations of many conductive metallic specimens the human eye cannot distinguish or identify. Many high grade specimens of different ores can be overlooked by the human eye for a variety of reasons, not the least of which is the fact that while the eye cannot see inside a piece of rock to look for valuable metallic specimens, a good quality VLF/TR metal detector can.

As previously discussed, "metal" refers to any metallic substance of a conductive nature in sufficient quantity to disturb the electromagnetic field of the searchcoil and produce a "positive" detector speaker and meter signal. Gold, silver, copper, and all the non-ferrous metals are just that — metals. "Mineral" refers mostly to minerals which cause a "negative" reaction in the metal detector — primarily magnetic iron and iron oxide ($Fe_3O_4$).

Conduct bench tests to familiarize yourself with the responses produced by various samples of which you already know the mineral content. This kind of testing will aid you greatly in testing future samples. Again, remember, use your detector as an *aid* in the field, not as a complete searching tool. Test any likely-looking rocks with your detector. You'll find that with just a few minute's work you might wind up with a high grade specimen that has been passed over for years by fellow rockhounds!

You will find that the VLF/TR detector will be the best to use in the identification of all metallic ores. The very low frequency of such a circuit penetrates the specimen easily, producing excellent results.

To identify metal *versus* mineral specimens, use the TR discriminating circuit rather than the VLF ground canceling mode. Set the discrimination control at ZERO.

(You must be sure your detector has its TR discriminate control factory set so that there is *no* rejection of *any* metal at the "zero" setting. See Chapter VIII.) This setting will serve to reject the specimens containing a higher content of magnetite. Now you may advance the discrimination setting slowly to determine the content of the specimen as to ferrous pyrite or non-ferrous precious metals, such as silver, gold, copper, *etc.*

Be very careful to test your specimen by moving it quickly toward and away from the *center* bull's eye of the searchcoil. Due to the winding configuration of the co-planar searchcoils, the sample may respond differently if tested on different areas of the searchcoil surface.

Thorough bench testing (a necessity if you are going to have good success with your detector) will quickly clear up any confusion that may exist in your mind. As an example, a large gold nugget placed on the *receiver* winding portion of the searchcoil will produce a metallic or positive response. Place the same nugget on the portion of the searchcoil containing the transmitting coil winding and it will then respond as "mineral," or negative. Place a small pebble of extremely high grade magnetite iron ore on the receiver coil winding and it will respond as "mineral." Place it on the transmitting portion of the coil winding and it will respond as "metallic" or positive.

To avoid confusion, conduct each test in precisely the same manner and use *only* the direct detection "center" (your bull's eye) of the searchcoil to obtain a correct reading. Used this way, the VLF/TR metal detector will result in the full enjoyment of your hobby and lessen the chances of missing a valuable specimen.

When searching for precious samples with your detector, pay close attention to old mine tailings. There could well be a high grade specimen of metallic ore around such a location and the metal detector used as your extra eye could identify it quickly. Become familiar with your detector! A whole new world will open up to you if you will just experiment. You will find you can spot-check promising rocks, and perhaps you will discover that those worked out areas aren't really so worked out after all.

Ask your metal detector dealer to demonstrate various marginal samples of easily recognizable gem or ore specimens. If the dealer does not understand identification procedures, suggest that he read a copy either of this book or THE COMPLETE VLF/TR METAL DETECTOR HANDBOOK, both published by Ram Publishing Company.

Do not lightly brush aside the possibilities offered the rockhound by the capabilities of ground canceling metal detectors. Many of the detectable metallics are worth a fortune at today's prices. As an example, Roy Lagal recently attended one of the larger gem shows where a lady rockhound was selling specimens from a well known gold mine. She had picked the specimens from the discards on the dump, broken them into baseball sized pieces, and priced them at $3 each. Upon examination, a few felt rather heavy, and Roy became curious as to the gold content. He asked permission to test them with his metal detector, and three of the specimens responded like almost solid metal. Roy explained to the lady that he thought the specimens contained a high percentage of pure gold. She replied, however, that she did not believe in metal detectors and that he could still have any of the samples he wanted for $3 each. Roy promptly bought the three specimens and had a rockhound friend saw them into slabs. Two of the three specimens were indeed heavily laden with gold, and some of the slabs later were purchased by a jewelry maker for $125 each. Roy has also high graded many silver samples that people had for sale. With a good VLF/TR metal detector it is a simple matter to pick the loaded ones. This is just one more way you can make your metal detector pay off!

# CHAPTER XIII

## *Conclusion*

Over the past several years we, the authors, have spent many long, hard hours in the laboratory and in field testing, retesting, evaluating, and using VLF/TR ground canceling metal detectors. We have proved many times just how valuable these detectors can be to the prospector and recreational miner. They can, certainly, spell the difference between success and a lack of it. The one basic ingredient of success is the operator's ability to use the versatile VLF/TR correctly. If you will study this book, search out the gold fields, and make a determined effort to apply in the field what you have learned herein . . . you cannot help but be successful.

We strongly encourage you to further your studies by reading THE COMPLETE VLF/TR METAL DETECTOR HANDBOOK and GOLD PANNING IS EASY. These books will round out your ability to understand and use the VLF/TR metal detector and the "Gravity Trap" gold pan, both very necessary tools.

The future of electronic prospecting is extremely promising. We wish for you many full pokes of the earth's treasures you seek. See you in the field!

*Charles Garrett*

*Roy Lagal*

# APPENDIX

## Where to Find Additional Information on Gold Areas

## STATE BUREAUS OF MINES AND GEOLOGY

Nearly every state in the Union has its Bureau of Mines and Geology. These offices are excellent sources of information on gold-producing areas in a particular state. Some of the addresses are listed below.

Alaska Division of
Geological Survey
3001 Porcupine Drive
Anchorage, AK 99504

Arizona Bureau of
Mines and Geology
University of Arizona
Tucson, AZ 85721

California Division of
Mines and Geology
1416 9th Street
Room 1341
Sacramento, CA 95814

Colorado Geological Survey
1845 Sherman Street
Room 254
Denver, CO 80203

Idaho Bureau of
Mines and Geology
University of Idaho
Moscow, ID 83843

Montana Bureau of
Mines and Geology
College of Mineral Science
and Technology
Butte, MT 59701

Nevada Bureau of
Mines and Geology
University of Nevada
Reno, NV 89507

New Mexico Bureau of
Mines and Mineral Resources
Campus Station
Socorro, NM 87801

Oregon Department of Geology
and Mineral Industries
1069 State Office Building
Portland, OR 97201

South Dakota Geological Survey
Science Center
University of South Dakota
Vermillion, SD 57069

Utah Geological and
Mineralogical Survey
University of Utah
Salt Lake City, UT 84112

Washington Division of
Mines and Geology
P. O. Box 168
Olympia, WA 98501

Geological Survey of Wyoming
P. O. Box 3008
University Station
Laramie, WY 82070

## WESTERN FOREST SERVICE REGIONS

To learn whether the area you intend to prospect is open for that activity, study land status maps to find out about open and closed areas. The U. S. Forest Service has land status maps for areas within the boundaries of a National Forest. The names and addresses of the seven Western Forest Service regions are shown below.

ALASKA REGION:
Federal Office Building
P. O. Box 1628
Juneau, AK 99802

CALIFORNIA REGION:
630 Sansome Street
San Francisco, CA 94111

INTERMOUNTAIN REGION:
324 - 25th Street
Ogden, UT 84401

NORTHERN REGION:
Federal Building
Missoula, MT 59807

PACIFIC NORTHWEST
REGION:
319 SW Pine Street
Portland, OR 97208

ROCKY MOUNTAIN
REGION:
11177 - 8th Avenue
Lakewood, CO 80225

SOUTHWESTERN
REGION:
517 Gold Avenue, SW
Albuquerque, NM 87102

## BUREAU OF LAND MANAGEMENT

The Bureau of Land Management (BLM) is under the U. S. Department of Interior. It controls all property outside of the National Forests. If you plan to do a lot of prospecting on these "public lands," or if you have located a good mineral deposit and wish to consider staking a claim, contact your nearest BLM office for additional information. The BLM also has maps showing which areas under their control are open to vehicle traffic, which may be prospected and which may not. The Bureau of Land Management also has a number of helpful brochures on mining laws. If you have any questions, drop them a line and ask for literature. Two bulletins you may want to start with are *Regulations Pertaining to Mining Claims Under the General Mining Laws of 1872* and *Staking a Mining Claim on Federal Lands* (Information Bulletin No. 4-76).

The addresses of the twelve BLM offices follow.

ALASKA STATE OFFICE
555 Cordova St.
Anchorage, AK 99501
(907) 277-1561

ARIZONA STATE OFFICE
2400 Valley Bank Center
Phoenix, AZ 85073
(602) 261-3873

CALIFORNIA STATE
OFFICE
Federal Building
Sacramento, CA 95825
(916) 484-4676

COLORADO STATE OFFICE
Colorado State Bank Building
Denver, CO 80202
(303) 837-4325

EASTERN STATES OFFICE
(All States bordering on and
East of Mississippi River)
7981 Eastern Avenue
Silver Springs, MD 20910
(301) 427-7500

IDAHO STATE OFFICE
Federal Building
Boise, ID 83724
(208) 384-1401

MONTANA STATE OFFICE
(Montana, North &
South Dakota)
Granite Tower Building
222 N. 32nd Street
Billings, MT 59101
(406) 657-6461

NEVADA STATE OFFICE
Federal Building
Reno, NV 59609
(702) 784-5451

NEW MEXICO STATE
OFFICE
(New Mexico, Oklahoma,
& Texas)
Federal Building
Santa Fe, NM 87501
(505) 988-6217

OREGON STATE OFFICE
(Oregon & Washington)
729 NE Oregon Street
Portland, OR 97208
(503) 234-4001

UTAH STATE OFFICE
University Club Building
136 E. South Temple Street
Salt Lake City, UT 84111
(801) 524-5311

WYOMING STATE OFFICE
(Wyoming, Nebraska,
& Kansas)
Federal Building
Cheyenne, WY 82001
(307) 778-2326

## OTHER SOURCES

Gold Prospectors Association of
America
P.O. Box 507
Bonsall, CA 92003
(714) 728-6620

Keene Engineering
9330 Corbin Avenue
Northridge, CA 91324
(213) 993-0411

Search International
P.O. Box 473007
Garland, TX 75041
(214) 278-6151

Miners, Inc.
Box 1301
Riggins, ID 83549
(208) 628-3865

*National Prospector's Gazette*
Segundo, CO 81070

For more information about treasure hunting

and related equipment call toll free–

1–800–527–4011 or 1–800–442–4889 (in Texas)

or write Search International ,

2814 National Dr., Garland, Texas 75041–2397 USA.

# A SELECTED READING LIST

Barlee, N. L. *The Guide to Gold Panning in British Columbia: Gold Regions, Methods of Mining and Other Data.* Canada West Publications, 1976.

_____. *Gold Creeks and Ghost Towns: East Kootenay, Boundary, West Kootenay, Okanagan and Similkameen.* Canada West Publications, 1976.

_____. *Historic Treasures and Lost Mines: Of British Columbia.* Canada West Publications, 1976.

_____. *Similkameen: The Pictograph Country.* Canada West Publications, 1978.

_____. Bureau of Mines State Liaison Officers. *Mining and Mineral Operations in the New England and Mid-Atlantic States: A Visitor Guide.* Washington: U. S. Government Printing Office, 1976.

_____. *Mining and Mineral Operations in the North-Central States: A Visitor Guide.* Washington: U. S. Government Printing Office, 1977.

_____. *Mining and Mineral Operations in the Pacific States: A Visitor Guide.* Washington: U. S. Government Printing Office, 1976.

_____. *Mining and Mineral Operations in the South Atlantic States: A Visitor Guide.* Washington: U. S. Government Printing Office, 1976.

Clark, William B. *Gold Districts of California.* (Bulletin 193.) Sacramento: California Division of Mines and Geology, 1976.

Dwyer, John N. *Summer Gold: A Camper's Guide to Amateur Prospecting.* New York: Charles Scribner's Sons, 1971.

Eissler, Manvel. *The Metallurgy of Gold.* London: Crosby Lockwood and Son, 1889.

Emmons, William Harvey. *Gold Deposits of the World — With a Section on Prospecting.* New York and London: McGraw-Hill Book Company, Inc., 1937.

Garrett, Charles, and Lagal, Roy. *The Complete VLF/TR Metal Detector Handbook: All About Ground Canceling Metal Detectors.* Dallas: Ram Publishing Company.

_____. *Modern Metal Detectors:* Home, Classroom, and Field Guidebook to understanding and using detectors. Ram Publishing Company, 1985.

Lagal, Roy. *Weekend Prospecting:* You can find an ounce of gold a day and the author tells you how, using basic equipment and techniques. Ram Publishing Company, 1985.

_____. *Gold Panning Is Easy.* 2d. ed., rev. Dallas: Ram Publishing Company, 1978.

Marx, Jenifer. *The Magic of Gold.* New York: Doubleday & Company, Inc. 1978.

Muns, George F. *How to Find and Identify the Valuable Metals.* Philadelphia and Ardmore, PA: Dorrance & Company, 1977.

Thornton, Matt. *Dredging for Gold . . . The Gold Divers' Handbook: An Illustrated Guide to the Hobby of Underwater Gold Prospecting.* Northridge, CA: Keene Industries, 1975.

# RECOMMENDED SUPPLEMENTARY BOOKS

The books described below are among the most popular books in print related to treasure hunting. If you desire to increase your skills in various aspects of treasure hunting, consider adding these volumes to your library.

**MODERN METAL DETECTORS.** Charles Garrett. Ram Publishing Company. NEW! This advanced handbook explains simply yet fully how to succeeed with your metal detector. Written for home, field, and classroom study, MMD provides the expertise you need for success in any metal detecting situation, hobby or professional. Easily understood chapters on specifications, components, capabilities, selecting and operating a detector, choosing searchcoils and accessories, and more — increase your understanding of the fascinating, rewarding fields of metal detector use. 544 pages. 56 illustrations, 150 photos,

**DETECTOR OWNER'S FIELD MANUAL.** Roy Lagal. Ram Publishing Company. Nowhere else will you find the detector operating instructions that Mr. Lagal has put into this book. He shows in detail how to treasure hunt, cache hunt, prospect, search for nuggets, black sand deposits ... in short, how to use your detector exactly as it should be used. Covers completely BFO-TR-VLF/TR types, P.I.'s, P.R.G.'s, P.I.P.'s, etc. Explains precious metals, minerals, ground conditions, and gives proof that treasure exists because it has been found and that more exists that you can find! Fully illustrated. 236 pages.

**ELECTRONIC PROSPECTING.** Charles Garrett, Bob Grant, Roy Lagal. Ram Publishing Company. A tremendous upswing in electronic prospecting for gold and other precious metals has recently occurred. High gold prices and unlimited capabilities of VLF/TR metal detectors have led to many fantastic discoveries. Gold is there to be found. If you have the desire to search for it and want to be successful, then this book will show you how to select (and use) from the many brands of VLF/TR's those that are correctly calibrated to produce accurate metal vs. mineral identification which is so vitally necessary in prospecting. Illustrated. 96 pages.

**GOLD PANNING IS EASY.** Roy Lagal. Ram Publishing Company. Roy Lagal proves it! He doesn't introduce a new method; he removes confusion surrounding old established methods. A refreshing NEW LOOK guaranteed to produce results with the "Gravity Trap" or any other pan. Special metal detector instructions that show you how to nugget shoot, find gold and silver veins, and check ore samples for precious metal. This HOW, WHERE and WHEN gold panning book is a must for everyone, beginner or professional! Fully illustrated. 112 pages.

77

**THE COMPLETE VLF-TR METAL DETECTOR HANDBOOK** (All About Ground Canceling Metal Detectors). Roy Lagal, Charles Garrett. Ram Publishing Company. The unparalleled capabilities of VLF/TR Ground Canceling metal detectors have made them the number one choice of treasure hunters and prospectors. From History, Theory, and Development to Coin, Cache, and Relic Hunting, as well as Prospecting, the authors have explained in detail the capabilities of VLF/TR detectors and how they are used. Learn the new ground canceling detectors for the greatest possible success. Illustrated. 200 pages.

**ROBERT MARX: QUEST FOR TREASURE.** R. F. Marx. Ram Publishing Company. The true story of the discovery and salvage of the Spanish treasure galleon, *Nuestra Señora de la Maravilla*, lost at sea, January 1656. She went to the bottom bearing millions in gold, silver and precious gems. Be there with the divers as they find coins and priceless artifacts over three centuries old. Join Marx's exciting adventure of underwater treasures found. The story of the *flotas*, dangers of life at sea, incredible finds ... all are there. Over 50 photos. 286 pages

**TREASURE HUNTER'S MANUAL #6.** Karl von Mueller. Ram Publishing Company. The original material in this book was written for the professional treasure hunter. Hundreds of copies were paid for in advance by professionals who knew the value of Karl's writing and wanted no delays in receiving their copies. The THM #6 completely describes full-time treasure hunting and explains the mysteries surrounding this intriguing and rewarding field of endeavor. You'll read this fascinating book several times. Each time you will discover you have gained greater in-depth knowledge. Thousands of ideas, tips, and other valuable information. Illustrated. 318 pages.

**TREASURE HUNTER'S MANUAL #7.** Karl von Mueller. Ram Publishing Company. The classic! The most complete, up-to-date guide to America's fastest growing activity, written by the old master of treasure hunting. This is *the* book that fully describes professional methods of RESEARCH, RECOVERY, and TREASURE DISPOSITION. Includes a full range of treasure hunting methods from research techniques to detector operation, from legality to gold dredging. Don't worry that this material overlaps THM #6 ... both of Karl's MANUALS are 100% different from each other but yet are crammed with information you should know about treasure hunting. Illustrated. 334 pages.

**SUCCESSFUL COIN HUNTING.** Charles Garrett. Ram Publishing Company. The best and most complete guide to successful coin hunting, this book explains fully the how's, where's, and when's of searching for coins and related objects. It also includes a complete explanation of how to select and use the various types of coin hunting metal detectors. Based on more than twenty years of actual in-the-field experience by the author, this volume contains a great amount of practical coin hunting information that will not be found elsewhere. Profusely illustrated with over 100 photographs. 248 pages.

**TREASURE HUNTING PAYS OFF!** Charles Garrett. Ram Publishing Company. This book will give you an excellent introduction to all facets of treasure hunting. It tells you how to begin and be successful in general treasure hunting; coin hunting; relic, cache, and bottle seeking; and prospecting. It describes the various kinds of metal/mineral detectors and tells you how to go about selecting the correct type for all kinds of searching. This is an excellent guidebook for the beginner, but yet contains tips and ideas for the experienced TH'er. Illustrated. 92 pages.

**PROFESSIONAL TREASURE HUNTER.** George Mroczkowski. Ram Publishing Company. Research is 90 percent of the success of any treasure hunting endeavor. You will become a better treasure hunter by learning how, through proper treasure hunting techniques and methods, George was able to find treasure sites, obtain permission to search (even from the U. S. Government), select and use the proper equipment, and then recover treasure in many instances. If treasure was not found, valuable clues and historical artifacts were located that made it worthwhile or kept the search alive. Profusely illustrated. 154 pages.

# BOOK ORDER BLANK

See your detector dealer or bookstore or send check or money order directly to Ram for prompt, postage paid shipping. If not completely satisfied return book(s) within 10 days for a full refund.

_____MODERN METAL DETECTORS $9.95

_____DETECTOR OWNER'S MANUAL $8.95

_____ELECTRONIC PROSPECTING $4.95

_____GOLD PANNING IS EASY $6.95

_____COMPLETE VLF-TR METAL DETECTOR HANDBOOK (THE) (ALL ABOUT GROUND CANCELING METAL DETECTORS) $8.95

_____ROBERT MARX. QUEST FOR TREASURE $11.95

_____TREASURE HUNTER'S MANUAL #6 $9.95

_____TREASURE HUNTER'S MANUAL #7 $9.95

_____SUCCESSFUL COIN HUNTING $8.95

_____TREASURE HUNTING PAYS OFF. $4.95

_____PROFESSIONAL TREASURE HUNTER $7.95

Please add 50¢ for each book ordered (to a maximum of $2) for handling charges.

| | |
|---|---|
| Total for Items | $ _____ |
| Texas Residents Add 6 1/8% State Tax | _____ |
| Handling Charge | _____ |
| Total of Above | $ _____ |

ENCLOSED IS MY CHECK OR MONEY ORDER $ _____

I prefer to purchase through my MasterCard (    ) or Visa (    ) account. (Check one.)

[MasterCard] [VISA]

_____    _____

Card Number                         Bank Identifier Number

_____    _____

Expiration Date                     Signature (Order must be signed.)

NAME _____

ADDRESS _____

CITY _____

STATE _____ ZIP _____

PLACE MY NAME ON YOUR MAILING LIST ☐

Ram Publishing Company

P.O. Drawer 38649, Dallas, Texas 75238

EP 7

214-278-8439

DEALER INQUIRIES WELCOME

81

## PUBLICATIONS

*THE SEARCHER:* This periodical is published and distributed by Search International, an association of metal detector enthusiasts. As a member, you are entitled to receive, free, each issue of THE SEARCHER, other publications and news bulletins as may be produced, and a 10% discount on all Ram books purchased through Search International. Each issue of THE SEARCHER is filled with treasure stories, treasure hunting and metal detecting "how to" information, and articles about found treasure by SEARCHER readers. Send name and address and $3.00 for one year's membership to SEARCH INTERNATIONAL, 2814 National Dr., Garland, TX 75041-2397. One year free membership with purchase of any Garrett metal detector from any Authorized Garrett dealer.